No Grid Survival Projects Bible

How to Manage Your First 1000 Days Off-Grid and DIY
Projects for a Thriving Homestead

Dale Mann

FREE GIFT TO OUR READER

Survival journal and video training

Please scan the QR code below to access the bonus materials. This will direct you to Google Drive. If you don't have Google Drive, please download it for free.

Table of Contents

Introduction

For those of you who have taken the journey before to be a prepper, you know that the prepper life is all about survival. But what if your DIY lifestyle was your choice?

Let's step beyond the canned goods that are stacked like Tetris blocks in your basement and the meticulously prepared bug-out bag that you have yet to carry. What if we ditched the quiet anxieties and the "what ifs," replacing them with "what if I could?" This book isn't about hunkering down and waiting for emergencies or for the world to collapse. This is a journey for those who've been pondering a less-ordinary path in life—a life where self-reliance isn't just a survival skill. Self-reliance here is about freedom and a chance to live like no one has before.

Imagine a sun-kissed morning where you aren't fighting with your early-morning alarms. Instead, the sounds you hear are the families of birds swooping by the window of your off-grid homestead. Think of how your pre-packaged breakfasts are replaced by things you have worked for: Fresh tomatoes from the vine, honey from hives you built yourself, and eggs from your chickens. To some, this would sound like an absurd dream, but there is an off-grid reality that is beyond the urban jungles and outside of suburbia.

Again, for those who have joined me before, we have dipped our toes into the pool of being a prepper. You've likely skimmed a few other survival manuals that are getting you ready for the end of the world. But let's be honest, many of the prepper journals are focused on something bad. And it's not that we shouldn't just ignore potential threats; that can't be the only reason we want to hone our crafts and be more self-reliant. This book aims to be something more—a blueprint for action that you can take right now. We aren't going to reinvent the wheel, but we will make sure that it's spinning smoothly.

When you go off-grid like this, you need to remember that this isn't a solitary journey. The off-grid life is much bigger than you think and still growing, which means there is so much shared knowledge out there for you. The off-grid community might be spread out, but all are brought together by the common goal of reconnecting to the earth, which means connecting. Throughout this book, you will find all those threads that have been made of those inspiring pioneers who embraced the off-grid path. Each project was passed down and enhanced by those who gave up the hustle and bustle for this tranquil life. Remember this as you work on your own projects: you are never alone.

For this journey, we will leave the doomsday preppers to their bunkers, jargon, and projects that need an entire aisle at Home Depot to complete. Let's think of ourselves as pioneers, the trailblazers who want a life that's less tethered and more connected to nature and the world around us. While we will dive into the importance of and why these chapters are included, we are going to skip the lectures and dive into practical, hands-on projects that you can do today with things that you might already have on hand. Think about turning your old toolbox into a catchment system for rainwater or building a solar oven with an old pizza box and some aluminum foil. Of course, we can't forget to turn your property into a miniature food forest with the right tools to make it sustainable.

Don't think of this as a collection of DIY projects. Think of it as your roadmap to self-sufficiency— something that points you toward that life that's about thriving instead of just surviving. Each chapter will act as a stepping stone that will take you through the fundamentals of off-grid living. We will cover things like water purification and storage (which those who joined my prepper book will be familiar with) and energy solutions, all the way to the basics of foraging. But we will also dive into the projects that will turn the empty land around you into an off-grid oasis. You will build, grow, raise, harvest, and secure your homestead with your hard work.

For those who did join in on our prepper journey, you can think of this as the annotated revision of the roll-up-your-sleeves-and-get-it-done DIY projects. This is a guide for those who are just starting in any type of survivalist lifestyle, and it's for those who read the first book and craved so much more. So yes, we will revisit some familiar concepts from the last book, but we will take it on with a fresh lens. The questions you'll be posing to yourself will be "How do I make this happen?" and "Can I do this with the materials I have on hand?"

This book is more than just surviving for 1000 days; it's about making your life more fulfilling. Each day will be more rewarding than the last, especially as you get accustomed to the lifestyle.

But this does come with a warning: this lifestyle isn't for the faint of heart. While you can work your way up to conventional comforts, it's not going to happen overnight. You have to be willing to go without, and you have to be willing to put in the work to get there. It won't all be easy, and it won't always be backbreaking labor. With that being said, if you are ready to get away from the tether of conventional life, then let's roll up our sleeves and get ready to spend 1000 days (and beyond) off-grid!

Chapter 1:

Embracing the 1000-Day Mindset

One thing you need to keep in mind when making this transition to an off-grid lifestyle isn't just about preparing for a doomsday scenario in a fortified shelter stocked to the brim with MREs and other rations. We aren't waiting for the end of the world here; we are embracing the concept of self-sufficiency in the world we currently live in. This is about shedding off the layers of dependence and creating your own life in connection with nature.

Let's take Robert, for example. Robert had been a lifelong disaster prepper, and his basement could have been mistaken for a small market. He would be stocked for years if everything went downhill. But when Robert visited a friend's off-grid farm, something changed. Of course, Robert understood the aspects of survival that came from being off-grid, but when he watched his friend and his friend's family living like they were still living in the suburbs, he realized that there was so much more. This was no longer about just survival; it was life—life with meaning.

While you will be eager to navigate your first 1000 days on your off-grid path, we have some things to cover first. We have to do this knowing that these thousand days can't be a rush job. This has to be a deliberate, step-by-step change. This chapter is going to equip you with something more than knowledge. It will equip you with the ever-important mindset to be an off-grid homesteader. You will gain the grit, adaptability, and resilience that are the foundation for this lifestyle, and then it will help you learn how to grow food, generate power, build shelter, and everything in between.

Imagine that this journey is just like the seed from your future gardens. Each day that you put effort into it, that seed will grow. Eventually, you will have an entire root system that spans your entire property, bringing you closer to the conventional life you're used to while still off the grid. We will start with the essentials, not to mention the array of practical DIY projects that you can implement as soon as you choose where you are going to create this homestead.

Remember that this journey won't be easy, and you will have challenges. Nature can sometimes be harsh, which means you will have to show respect and constantly keep learning about what's happening in your surrounding area. There will be doubts and setbacks, and they won't be easy to face. However, it takes these things to build the foundation of your homestead.

Every obstacle that you overcome, every lesson learned, and each harvest you bring in will strengthen your resolve and reinvigorate your passion. So, take a deep breath and start to let go of the prepper mentality. Then take another and embrace the 1000-day mindset.

1.1 Redefining Survival

The prepper mentality is pretty well known. It focuses on impending disasters and stockpiling supplies for when the world collapses, which makes it feel more like an act of anxiety. Not that it's a bad thing because the future is uncertain. But what if we changed our frame of mind? What if, instead of hunkering down when the worst does happen, we embraced a different kind of preparedness? What if we rooted ourselves in resilience, resourcefulness, and a deeper connection to the nature that surrounds us? This is where the world of off-grid homesteading comes in. This is less "doomsday prophet" and more "builder of the future." It's about choosing a life that thrives on its own terms.

Preppers vs. Off-Gridders: A Shift in Mindset

Again, the prepper mindset hinges on that one cataclysmic event—war, societal collapse, major natural disasters. It's about waiting for the world to end (even if it's the world around them) and doing what they can to survive in the aftermath. The off-grid homesteader, however, is a master of adaptation. They can see that there are inevitable challenges, but instead of approaching them from a place of anxiety, they see them as a chance to learn, grow, and build something better—something sustainable. Off-gridders are not focused on a disaster but on making a life that is resilient to anything. They do it because they want to be ahead of the curve.

You can see this mindset shift in how they approach preparedness. Think about your normal prepper. They usually stock up on non-perishable items, guns and ammo, survival gear, and other things, and they are making plans for events that could or could not happen. The off-grid homesteader, on the other hand, is more invested in learning skills and accumulating knowledge. They want to stick to the DIY lifestyle and do things like grow their food, generate their energy, and build things themselves. This form of self-reliance is driven by the desire to be empowered and to have complete control over their own lives.

Again, there is nothing against the prepper. Those are great skills to have, and we should still be prepared for something to happen. But off-grid takes a big step back from that narrow scope and thinks about more than a disaster scenario. Instead of a bug-out bag, you're either building a house or retrofitting your current one. Instead of three days' worth of food, you're thinking about what vegetables you're going to grow for the next year. This is about building a life that is rich, fulfilling, and independent, whether there is a disaster or not.

Reclaiming Control and Responsibility: The Power of Off-Gridding

So, by now, you likely have a solid understanding that the decision to go off-grid is about reclaiming control. It's the awareness that you have been reliant on municipal systems for so long, but now you are shedding that layer of dependence. You don't want to rely on external systems. Instead, you want to be more connected to the land and yourself.

So what other empowering benefits await you when you make the shift?

- **Well-being by your design:** When you take full ownership of your energy, food, and water, you are an active participant in your well-being. The longer you're off-grid, the deeper your understanding of yourself is. You will get a better idea of what you need and the resources you have to meet them. This will foster a sense of responsibility and empowerment that we rarely see in our society today, as everything is driven by consumerism.

- **A problem-solver's mindset:** This is not a walk in the park, though. The off-grid lifestyle is riddled with challenges. One day, you might be fixing a leaky roof, and the next day, you're repairing your tools. And that doesn't even include determining how you will preserve your harvest. You have to constantly be resourceful and innovative, and that attitude will start to spill over into all aspects of your life. This is a great way to build confidence and self-reliance, which are invaluable traits in any situation.

- **Confidence in the face of uncertainty:** The prepper mentality does give you an understanding of one thing, though, that carries over to off-gridding, and that is that the world is a very fragile place now. We are staring in the face of climate change, political tensions, and an unstable economy. When we embrace the off-grid life, we aren't distancing ourselves from society and putting up a barrier; we are simply building our resilience. When you master essential skills and connect with your surroundings, you are less vulnerable to external devastation and more confident to navigate any of life's challenges.

Again, you are not retreating from the world, but you are actively building a better one. Think of this as the more advanced stage of preparation. It's living your life the way you want, with or without a disaster.

1.2 Conquering Initial Challenges

Stepping from the well-worn paths and embracing an off-grid lifestyle isn't just about leaving the suburbs and streetlights for starry skies. This is a monumental shift into brand-new territories—ones filled with so much potential as well as a lot of challenges. This is a change that needs careful planning and some practical skills, but you are also going to need the unwavering spirit to navigate the internal challenges that will come and the inevitable pushback from external sources.

Challenges of Self: Taming the Doubts and Fears

Even as you think about your new life out there in the tranquil embrace of nature, you might run into some anxious thoughts and questions. It's common to ask yourself things like, "Is this the right choice?" or "Am I ready to handle this much independence?" There are others, but they are all normal doubts, and they are probably necessary. You should acknowledge them, understand them, and accept that they are there. Don't let the whispers of doubt become roadblocks; instead, see them as a way to self-discovery.

When these moments of uncertainty happen, combat them with introspection. Ask yourself what is drawing you to an off-grid lifestyle. Think about what your goals are and what your core values are. From there, dive into the research. Besides this book, soak in as much information given by off-gridders as you can get. Seek out communities online, where you'll find people who have carved their paths and will give you guidance and encouragement. These avenues might even lead you to local connections. The important thing is to remember that you're not alone in this. There is a tribe of off-grid homesteaders that are walking with you.

One of the biggest internal challenges will come from the financial side of things. In all honesty, the transition to an off-grid life is going to require a lot of rethinking when it comes to your money. You have to think about your income and adopt more frugal habits. To combat this, create a realistic budget that will help you account for every cent you spend. This should include everything from land acquisition to your energy solutions. Look at other sources of income like remote work opportunities, crafting skills, or small-scale agriculture, which are all ideal for an off-grid lifestyle. Also, you don't have to have accumulated wealth to make this change; you just need to make sure you have the means to get started and live comfortably.

External Challenges: Combating the Doubts of Others

Not everyone is going to understand why you chose an off-grid lifestyle. Most of the time, your family and friends will mean well, but they will likely raise concerns, voice skepticism, or even try to make you envision a bleak future. It's easy to get caught up in this negativity and question yourself, as well-natured as the sentiments are. This is when you'll need to stay grounded in your vision. Acknowledge their worries and even show gratitude for their concerns, but then communicate your resolve and your reasoning for this lifestyle change.

Resilience against that external negativity is going to be so crucial. You can't let their doubts become your own. Again, they mean well, but you should surround yourself with a supportive network filled with people who resonate with your vision, believe in your ability to do it, and celebrate every milestone and victory with you. Look for mentors and off-grid communities, and create bonds with other off-grid homesteaders. It will be a great space to share stories, advice, and triumphs. Your network will also keep you motivated and on track.

Ultimately, conquering internal and external challenges will be an ongoing process. You will never be able to completely escape the voices of doubt, but you will need to learn how to navigate them with courage and self-compassion. Each hurdle you overcome will strengthen your resolve. Your off-grid journey is a big step in a bold direction, but after you've determined that this is the life for you, nothing can stop you.

1.3 The Pros and Cons of Off-Grid Living

Some will see only the negatives of off-grid living, and there will be some who only see the positives. Only seeing one side of the coin will sway a person's decision, and many times the person will feel like they made the wrong decision. For example, a person who only sees the good things might feel overwhelmed by the challenges this lifestyle holds. So the other way that you will build your mindset is to weigh out the pros and cons.

The Pros: Stepping Off the Grid and Taking Back Control

1. A Life Without Ties

Imagine if you could just shed the ties of dependence that you have on conventional utilities, the constant pressures of debt, or the endless cycle that is the "rat race." Off-grid living gives you a path to independence where your livelihood isn't dictated by external forces. Instead, you're generating your energy, growing your food, and managing your resources. It's through this that you can foster a profound sense of self-reliance and empowerment.

2. Embracing Nature

Homesteading lets you step away from the urban and suburban jungles and step into a world where things are done at your own pace, aligning with the rhythm of nature. When you're out on your homestead, you can immerse yourself in the tranquility of nature. Even if you power your home to the luxuries you're used to, you always have the option to step back into the quiet calm. This reconnection with the natural world will give you a sense of peace and purpose and a deeper appreciation for the simpler things in life.

3. Prioritizing What Matters

When you choose to live off-grid, you are making the conscious decision to prioritize things that truly matter. It's a big shift away from the pursuit of material objects to building deep, meaningful connections with family, community, and your well-being. Choosing the off-grid life encourages a slower pace, mindful consumption, and a focus on experiences instead of possessions. This leads you towards a richer and more fulfilling life.

4. Unconventional Learning

Living off-grid turns into a lifelong learning experience. You'll learn new skills while mastering others, from carpentry and your energy systems to food preservation and natural medicine. When you are constantly learning, you are engaged. This will keep your problem-solving abilities sharp, and it will foster your sense of accomplishment.

5. Environmental Stewardship

Being off-grid means embracing renewable energy sources, minimizing waste, and living sustainably. Because of that, you become an active participant in protecting our environment. As more people disconnect from the grid, there is a positive impact on our planet, ensuring that it's healthy for generations to come.

The Cons: Challenges and Considerations

1. An Investment in Labor

Off-grid living is not for everyone, and even skilled preppers can struggle. Life off-grid requires a significant investment of time, physical effort, and a lot of mental resilience. There are going to be some very long days, intensive labor, and a constant need to troubleshoot and maintain your systems. This lifestyle demands a strong work ethic and a willingness to get down and dirty.

2. Limited Amenities

When you trade a convenient life for self-sufficiency, you'll be sacrificing some of your modern comforts. You are now further away from medical care and specialized services, and even your entertainment options may be limited. This is going to require you to embrace improvisation and resourcefulness, especially in the beginning, as you adapt to the simpler way of life.

3. Financial Investment

Setting up your off-grid homestead is going to require a substantial amount of money upfront as you invest in land, infrastructure, and equipment. The savings are there, but you will see them over time. It's the initial costs that can be quite daunting, and this is why careful planning, budgeting, and finding other financing options are so important in your journey.

4. Social Isolation

The off-grid lifestyle can lead to long periods of isolation, especially the more remote your homestead is. If you are in a rural setting, this is where building strong connections with your community and potential off-grid neighbors is necessary. If there are people near you, they will be essential for your emotional well-being and support. You might have to embrace alternative forms of communication and find other opportunities for social interaction.

5. Uncertainty and Adaptation

Off-grid living comes with many uncertainties. Fluctuations in weather can impact your energy production, equipment can (and will) malfunction, and there are a myriad of unexpected challenges. You have to be ready at all times to adapt, improvise, and troubleshoot with a calm and resourceful mindset.

***

Ultimately, the decision to hop off-grid is a very personal choice. You have to weigh all your pros and cons carefully, and you have to account for your individual needs and resources. This journey also needs to be made with realistic expectations and a spirit of adventure. This is more than leaving the grid behind; it's about creating a new life that aligns with your values and enhances all of your connections.

1.4 The Importance of Mastering Your DIY Skills

After you've weighed the pros and cons, your mind wanders back to the dream of trading the business of modern life for the quiet life of your rural or remote homestead. But we have to stop one more time because there is still something that you have to think about. We've hinted at it, but let's dive into the ever-crucial skill of becoming a jack-of-all-trades.

Becoming a Jack-of-All-Trades: The Bedrock of an Off-Grid Life

The skills that you probably brushed off as "handyman stuff" are going to become your daily companions. Carpentry skills move from weekend checklists to building shelters, livestock pens, or even furniture. Plumbing is suddenly a matter of maintaining your water systems, such as fixing leaky faucets or potentially creating a makeshift irrigation setup. Basic mechanic skills reach beyond an oil change for your car; you are now in charge of maintaining your tools, generators, and all the other machinery that keeps the homestead lively. Oh, and let's not forget first aid, which you know from being a prepper, is a must-have skill in any remote location where medical help could be miles away.

However, mastering these skills is only the first step. From there, you have to cultivate a growth mindset—one that thrives on challenges and sees setbacks as a way to get better. You should embrace the thrill of learning something new, whether you're deciphering a complex wiring diagram with your solar array or tackling the intricacies of agriculture. The journey to being a true jack-of-all-trades is a marathon race, not a sprint, which means you will never stop learning. Instead of feeling the frustration of not knowing everything right away, celebrate your small victories and learn from your mistakes.

Resourcefulness and problem-solving are the secret weapons of every homesteader. You will use these to transform limitations into opportunities. If you have a broken tool, you don't just stop the project. Instead, you jump at the chance to improvise. You could use your ingenuity to repair the tool, or you could look for a different solution. You are using the "make-do with what you have" spirit, which is the ability to see beyond limitations and find a creative solution for any problem.

Beyond the Basics: A Brief Overview of Skills Needed When Off-Grid

Now, we can dive deeper into a rough idea of the specific skillset you'll need to build when you establish your homestead. While you want to focus on the "jack-of-all-trades" aspect of your skills, these areas will require more of your time.

- **Sustainable energy systems:** You will be required to understand, maintain, and potentially install renewable energy sources like solar panels, wind turbines, or micro-hydro systems. You should know deeply about battery storage and efficient energy use when powering your home.

- **Water management:** You will have to work with your climate, so you'll need to learn as much as you can about rainwater harvesting, well maintenance, irrigation techniques, and basic water conservation. Management of the most precious resource is essential to your homestead.

- **Waste disposal:** Composting, greywater systems, and other responsible waste disposal are crucial for keeping your environmental impact to a minimum. It also does wonders for maintaining a healthy off-grid ecosystem.

- **Basic appliance and infrastructure repair:** You will have to know how to fix a wide array of minor issues with appliances, tools, and things like plumbing or electrical systems. Remember, preventative maintenance can save you time and resources down the line.

The important thing to remember is that being an off-grid homesteader isn't just about skills; you are embracing a mindset of simplicity. Do what you can to live with less and find happiness in being resourceful.

Now, you have more in-depth knowledge about what to expect and what it means to be an off-grid homesteader. It's vastly different than being a survival prepper, but with resolve and embracing the growth mindset, you will thrive in any location where you choose to establish your homestead. Speaking of which, we are now in the first part of the journey. You can't really thrive off-grid for 1,000 days if you don't have a homestead established.

Chapter 2:

Setting Your Homestead

The mindset is not difficult to master, and many will find themselves ready to step foot into the unknown. However, imagining the perfect life as an off-grid homesteader is one thing; getting out there is another. One of the first challenges that you will face is finding where your homestead is going to be. Not only do you want the house to be comfortable, but you also have to look closely at the very foundation of your new life: the land itself.

Tim was a former city dweller who had the same desire to go off-grid, but he had in his head that off-gridding meant that he needed to be far away from people. He wanted a balance of convenience and a connection with nature. He searched for months until he found a small farm nestled in the hills, just outside of a small town. The land had everything he could have hoped for. There was fertile soil, a stream that flowed through the property, and even a couple of outbuildings that he could fix up. With that and a town that was 15 minutes away, Rob had found his Eden.

Whether you want to be close or far from convenience, finding your own off-grid Eden will take careful consideration. This is so much more than gorgeous views and peace and quiet. In this chapter, we will get into the nitty-gritty of finding your off-grid homestead. You will know what to consider when choosing a site, building the home itself, what to consider about the surrounding land, and a brief introduction to powering your homestead. Remember that in order to have a successful homestead, you have to be able to work with the land around you.

2.1 Finding Your "Eden"

Through the last chapter, you learned that choosing the off-grid lifestyle is a bold choice, filled with as many challenges as there are rewards. This brings us to our first challenge, the crucial step that you have to do before you get down to any DIY projects: choosing where you will establish your homestead.

Choosing this haven is so much more than just finding a plot of land; it's about choosing the very foundation of your off-grid homestead. When you think about the first 1000 days of being off-grid, this is the most important time because you establish roots and build resilience. And while there are challenges, you don't want it to all feel like an uphill climb. So, let's look at some of the most important factors to consider when finding your "home-sweet-home."

Climate and Accessibility

A homesteader is looking for harmony with nature, which is important, but you must remember that nature can be harsh. Think about your desired lifestyle and your agricultural goals and choose a place with a climate that suits them.

It would help if you also thought about accessibility, with year-round access being essential. Imagine choosing somewhere where you're snowed in all winter with no access to essential supplies or services. Another important consideration is reaching your property during extreme weather, which might be common in the area. Therefore, when considering an area, research the average snowfall, flood risks, and road conditions around the property to ensure that you can get through every season and the worst weather.

Resource Considerations

You might be tempted to choose an area based on looks alone, but you have to think beyond a picturesque landscape. You have to think about harnessing the resources that your chosen location has available. You need resources to be successful, and any limitations in resources will greatly increase the difficulty of your off-grid life.

Think of all the things that are going to be important to your homestead. Fertile soil is needed if you want to have a better chance at a productive garden. You also know that water is the lifeblood of your survival, and your homestead should have plenty of access to it. That means you have to think beyond just rainwater catchment systems; being near a natural source like a stream or a river is great, or you will need to have a well installed.

Don't overlook other potential resources. You could use timber for construction, plants for medicinal purposes, or clay deposits to make pottery or earthenware.

Sun, Slope, and Sustainability

Sunlight will be one of your greatest allies, as you will need it for growing food and harnessing solar energy. Make places that have abundant sunlight exposure a priority. This will ensure that you not only get sufficient energy production but also optimal growing conditions.

The topography is going to also play a role, especially when you're considering your agricultural needs. A gentle slope is tremendous when you want to create microclimates on your land and make the most of your rainfall. South-facing slopes will capture the warmth of the sun, which is great for plants that love heat. If you have a north-facing slope, you will have a perfect climate for leafy greens.

Zoning and Regulations

Your self-sufficient dreams can hit a major wall or fall apart completely because of legalities. After selecting a potential location, take time to familiarize yourself with the local zoning laws and building codes. From there, you will need to step beyond the local realm and research state and federal regulations that will impact your plans.

These impacts can include restrictions on water usage, renewable energy systems, and even specific building materials. This can seem a little discouraging, but understanding these regulations empowers you to plan effectively and avoid taking steps that could be costly later. Also, keep in mind that some resources are going to be off-limits because of legal restrictions. An example of this is being prohibited by state and federal laws from using river water for your homestead.

***

When you consider these key factors (climate, resources, sun exposure, topography, and regulations), you will set your homestead up to thrive. Remember, your first 1,000 days off-grid begin with the right plot of land. With the right area, your journey will last well beyond the first 1,000 days.

2.2 Building the Off-Grid Homestead

Once you have a location selected, you will now need to shape the place where you will rest and recharge—the place you'll call home. This is the very heart of your homestead, which means it deserves thoughtful planning that makes comfort and sustainability top priorities. Your home can come in many forms. You could be renovating an existing structure, building from scratch, or preparing to have a prefabricated home put in. No matter your choice, let's look at some key considerations that need to go into your off-grid sanctuary.

Choose the Spirit of Sustainability

Sustainability should be the principle that guides you. Prioritize natural and recycled materials, which can include wood, straw bales, salvaged lumber, or reclaimed stone. These give you plenty of environmental benefits, but they require less maintenance. They also blend well with the natural landscape.

Keep in mind that there is no rule out there that forbids you from using commercial materials. If you opt for a prefabricated home, then that's fine. Any commercial material can be used when necessary, as long as you try to seek out options like recycled materials or those with low environmental impact.

Using the Power of Passive Design

Try to use a passive solar design whenever possible. This design is going to do wonders for the home as strategically placed windows become natural heaters and light sources, which minimizes your reliance on external energy. This is helpful in the winter months as the days get shorter and when you need to divert energy to other needs in the home.

South-facing windows are going to warm your place wonderfully during the months that matter, while strategically placed overhangs can regulate the summer heat. Beyond the sunlight, think about using insulation made from natural materials like wool, cellulose, or cork, which will retain heat in the winter and cool air in the summer.

Supplementing Your Water with Rainwater Harvesting

Rainwater harvesting is going to be an absolute game-changer for your off-grid lifestyle. Installing the most basic system will allow you to collect rainwater for various uses, from watering your garden to flushing your toilets. Rainwater systems will also reduce your dependence on external water sources, which helps conserve them. We will get back into this in a bit, so we won't get deep into the technicalities, but remember, a rainwater harvesting system is a needed component of your off-grid home.

Conventional Living with Off-Grid Energy

Off-grid living and generating your power go hand-in-hand. Again, we won't go into the fine details just yet, but this is a consideration that you will need to make. While some take a more traditional pioneer way of life with minimal energy, some homesteaders have all of their modern conveniences thanks to energy generated from natural sources.

The go-to method is through a solar array and harvesting energy from the sun. Some homesteaders are in windy areas, so they will opt to use wind turbines or a combination of the two. Some homes will even utilize hydropower systems from the flowing sources that run on their property.

It's important to research all options before choosing, though. For example, you might live near a water source, so hydropower might seem like the way to go, but you will likely need permits to use the water. Just make sure you are choosing the right option for your specific location.

Building the Dream

Before we move on to the land around your home, we should touch base again on the structure that will be your home. What type is best for you? What are the things that you want to know?

First, there are three things you want to remember.

- **Location is paramount:** Choose a site that offers you good sun exposure, wind potential (if you are going to implement wind turbines), and access to water sources.

- **Mind the regulations:** Make sure you know local building codes, other regulations, and zoning laws before building. There may be an additional cost for permits for off-grid structures and other construction considerations.

- **Build a network:** Connect with other off-grid homesteaders for support, advice, and shared experiences.

So now, let's look at your structure options again. There is no right or wrong answer to what you choose. As you know, the thing that matters most is sustainability.

- **From scratch:** Building your homestead from scratch gives you complete control of your home's design and the materials that you'll use. This is one of the more rewarding but more challenging routes. Going with this option requires careful planning, a good handle on construction skills, and a large time investment. If you use modular construction, this will be a faster, more manageable approach.

- **Renovation:** Transforming an existing structure into your home can be very budget-friendly and time-efficient. You want to find buildings with a solid foundation and a good framework, which usually means they need some energy-efficient upgrades. These structures can sometimes hide some huge challenges, so factor in time delays and budget adjustments.

- **Prefabricated homes:** The prefab market has ramped up, and the homes now come in many different designs; you can even find some that are built with sustainable materials. These are great for those who might not have the construction know-how at the beginning of their off-grid journey. They are also cost-effective, but you might be limited in the customization department.

2.3 The Land Around the Home

When you establish the heartbeat of your off-grid homestead, you have to look beyond that. You have to look at the very ground beneath your feet and nurture the land that is going to sustain you. Then, you have to learn and implement the basics of growing and raising your food. But the work doesn't stop there. You finally have to master the art of preservation after your harvest. The land around your home isn't going to provide you with just enough; it should provide you and yours when the cold winds settle in, and the days are short.

Again, we won't do a deep dive yet, as I am just introducing these concepts to you.

Soil Health and Composting

You want the land around you to be teeming with plant life, and that's why it's so crucial to have healthy soil. The fertile ground will provide the foundation for what will keep your homestead operational. It will provide the needed sustenance for your gardens and ensure that they have long-term viability. Learning how to maintain and improve soil health is crucial before planting a single seed.

Composting is going to be your biggest ally. This is the art of turning your kitchen scraps and yard waste into nutrient-rich "gold." Compost puts nutrients back into the soil, helps retain moisture, and promotes beneficial microorganisms, which creates a garden ecosystem that thrives every growing season.

There is more to cover like cover-cropping and crop rotation that will help keep your soil healthy, but we will cover this later.

Beginner Gardening Skills

Establishing a homestead is a step into the world of gardening, where you are creating and working with nature. However, before you break ground, you have to have an understanding of your local hardiness zone. This map categorizes regions by their average temperatures in the winter. It will help you choose the plants that are going to thrive in your climate.

The best way to start your garden is small, with easy-to-grow vegetables and herbs like lettuce, tomatoes, and basil. This is why homesteaders will work with this before making the transition off-grid. It allows you to overcome the stumbles of beginning gardening when your crops matter most. Not that there won't be setbacks and the need to produce the way you're used to, but you want to do as little of this as possible.

By starting now, you'll gain confidence, which means you can experiment with different varieties of vegetables as you also work toward an extended harvesting season with cool-weather crops.

Exploring Beginner Livestock

Some homesteaders venture beyond the plants and open their land up to raising animals for meat, milk, and eggs. This is a step that does require a good chunk of land, especially with a lot of animals or large ones; therefore, not everyone can implement it. If you do have the land, though, you will need to give it careful consideration.

Think about what animals you'd want to raise, and then research the needs of each animal, from housing and feed requirements to social behavior and even health concerns. By bringing animals into the homestead, you will need to be ready to dedicate time and resources to taking care of them. It's a very significant responsibility, but the rewards are going to be tremendous.

Preserving the Bounty

Preserving food is a great skill to learn as a prepper, but it is also needed as an off-grid homesteader. The harvest can't end as the seasons change because you need food during the non-growing season. Learning how to preserve your harvest will ensure you can enjoy it throughout those months and even the year. You'll want to master skills like canning, freezing, drying, and fermenting, which will all extend the shelf life of your produce, reduce food waste, and keep your pantry stocked with all your favorite foods.

Experiment using different methods to find what fits best with your taste and lifestyle. If you have more than enough, then you can sell your excess for extra income or trade with an off-grid neighbor for some of their goods.

2.4 Powering the Homestead

When you choose the off-grid lifestyle, there is nothing quite like reconnecting with nature and soaking in the self-sufficient lifestyle. However, as enjoyable as the sounds of birds and running water can be, some modern comforts can go a long way when taking on your first 1000 days and beyond. Powering your home with renewable energy sources will let you enjoy things like lighting, refrigeration, communication, and even entertainment while still being in harmony with the environment. The beauty is that your level of connectedness is completely up to you. We will get into this later on in the book, but let's introduce those concepts again.

Harnessing the Sun's Bounty With Solar Power

Many off-grid households don't just enjoy the warmth from the sun's rays; they also love taking that energy into powering their appliances. Solar panels will convert sunlight into energy, which makes them a clean and silent source of power. This setup is perfect for sunny regions, and you can make a small-scale array for basic needs, or you can go larger to meet more demands.

The only aspect that catches many off-guard is the initial investment. While it bears a significant price tag, government incentives and long-term energy savings will make the investment worthwhile.

Capturing the Breeze With Wind Power

Wind energy can be a great ally for your homestead. The same forces that blow across your property can be harnessed to generate electricity through wind turbines. Imagine a small windmill spinning atop your home, an animal shelter, or various structures.

This setup is ideal for homesteads that get a constant breeze. With the right location, you're looking at an extremely reliable and sustainable solution. While large turbines will be a more efficient solution, not everyone will have the means to implement a larger one. However, smaller models are still going to contribute significantly to your energy needs. Homesteaders will even use wind energy to power items around the property while using another renewable source (like the sun) for their homes.

Going With the Flow by Using Hydropower

If you have a babbling brook or a steady stream running through your property, you should consider channeling the power of that water. Hydropower systems will use turbines to convert the kinetic energy of the flowing water into electricity. It's like the wind, except with water. You can also have a small river system, or you can opt for a large dam; the scale of your system should match the resource you're using.

The drawback to these systems is that they require responsible planning and being mindful of regulations. While the source might be on or by your property, there may be regulations in place about using it as an energy source. Just be mindful of these to ensure you're operating in a legal capacity and making a minimal environmental impact.

Turning Waste into Energy With Biogas Digesters

You might not have considered it before, but your organic waste, like food scraps and manure, can be transformed into a potent fuel source. Biogas digesters create a controlled environment where microorganisms break this organic waste down, producing biogas (a mix of methane and other gases). You can then burn this gas to generate electricity, heat the home, and cook your meals. This solution has become popular for heating and cooking needs because not only does it provide clean energy, but it also minimizes waste while promoting nutrient recycling. While this does take some work to master, you have a truly sustainable solution on your hands that can save your other energies to maintain conventional living while off-grid.

2.5 Embracing Resourcefulness

The off-grid life calls to many because of the simplicity of the lifestyle and the bonding with nature. But as I introduced in the last chapter, venturing out beyond conventional life demands a unique set of skills. Self-reliance out here isn't just a cool attribute to have; it's the cornerstone of survival and success. Again, these are concepts that have been introduced to you already, but as we prepare to dive into our DIY projects in the next chapter, we should take one final look at the skills that are needed when approaching the off-grid lifestyle.

Basic Carpentry and Plumbing

Your off-grid home is going to be your castle, and basic carpentry skills are needed to build and maintain it. There will be times when you are called on to fix a leaky faucet, or you might need to make some new, sturdy furniture. When you learn and hone your carpentry skills, you will be able to solve common issues without needing to rely on external help. In that same frame of mind, you should learn how to use hand tools effectively. You should still invest in some power tools for larger and more intricate projects, but never underestimate basic skills.

Plumbing knowledge will also be crucial, especially when you are miles away from any type of emergency service. Knowing how to fix a burst pipe or install a new sink will be a breeze with the right amount of knowledge. These skills won't just help you save money; they will also enhance your sense of accomplishment and self-reliance.

Mechanical Competency

Your tools and appliances need to be trusted companions when you're living off-grid. You will need to get some basic engine maintenance knowledge to keep up with things like your tractor, generators, and other equipment. This will ensure that everything runs efficiently and will last you for years. You also should know how to troubleshoot common issues and perform simple upkeep like changing the oil. The most basic inspections can prevent costly problems down the line. When you gain mechanical competency, you will not only expand the life of your machinery but also be empowered to face those unexpected breakdowns with confidence.

Wilderness Medicine

Again, you're living away from external help, which includes medical facilities and first responders. Having a deep understanding of first-aid and wilderness medicine becomes a very crucial skill for your first 1,000 days and beyond.

Much like preppers, you should have a deep knowledge of essential skills like wound care, CPR, and emergency treatment for the most common injuries and illnesses. Your homestead first-aid kit should be tailored to your needs and location. For example, having supplies for bigger emergencies is important when living well outside the immediate reach of medical assistance. Taking first-aid courses is also a great idea for any situation because that basic knowledge could mean the difference between life and death.

Resourcefulness and Improvisation

One of the greatest realities of off-grid life is that you are dealing with limited resources. As a resourceful person, you should be defined by your ability to think creatively and adapt to any challenge laid before you. This means you should learn to identify and utilize alternative materials that are readily available in your surroundings.

Think about it. Should a broken tool mean that you always need to stop in the middle of your workflow to drive back to the city? Maybe it can be fixed with a repurposed piece of work or something that comes from your creative mind. Don't be afraid to improvise, and keep in mind that the simplest solutions are sometimes the most effective.

***

All the skills we mentioned in that last section are powerful tools, but remember that the true essence of self-reliance is in adopting an off-grid mindset. As long as you are willing to learn, experiment, and think outside the box, you will push through any challenges that your off-grid journey will present to you.

With all that said over the last couple of chapters, and with all these concepts now embedded in our heads, we can start opening the door to DIY concepts and projects for your new homestead. To kick off the new DIY journey, we will start with the most important element of them all: water, the lifeblood of your off-grid homestead.

Chapter 3:

Water—The Lifeblood of Survival

The earth was parched, and it stretched out for miles, a cracked canvas of despair. Cape Town, South Africa, was a vibrant town, but it had become a victim of its prosperity. The population had boomed, but this led to them stripping away at the dwindling water resources. Rainfall had also been scarce for over three years, which only contributed to the problem. The year was 2018, and Cape Town had been put under a dark cloud known as "Day Zero." This was the day that the water sources were predicted to run dry. (Heggie, 2021)

This warning led to panic buying and heavy water restrictions. The people were in despair, but innovation took over amidst that separation and the dust. Communities in Cape Town rallied together, implementing rainwater harvesting systems on their rooftops. Greywater reuse became a staple of everyday life, and desalination plants buzzed to life, taking the precious resource from the ocean, and giving it back to the people. Cape Town saw Day Zero come and go without seeing the prophecy of no water come true. They prevailed. They were put through a huge test, and they came out much more resilient and with a newfound respect for the element that is crucial to all life—water.

This story, which is now carved into the walls of our recent history, serves as a powerful reminder of just how vulnerable we are and the urgent need to find sustainable water solutions.

But what if we did this—not because we had to, but because we made the deliberate choice? Part of your 1,000-day journey is going to involve you severing your ties with conventional water grids and stockpiles of jugs and bottles. This is where you will get a real taste of how this life isn't for the doomsday prepper, but it's for the intrepid spirit, the person who wants to look beyond conventional life toward self-sufficiency.

Can anyone create a place independent of the ever-growing need for water?

The answer you had should have been a resounding "yes." However, that path is not paved with filters and plastic barrels. You are going to embark on another journey of exploration, careful planning, and a deep commitment. This is about understanding the surrounding land, harnessing the cycles of nature, and tapping into wisdom from long ago. We will delve into a world of water harvesting techniques that will benefit you for the first 1,000 days and into the future.

3.1 Identifying Your Main Water Source

Trying to secure a reliable source of water is the most crucial aspect of your off-grid homestead. This chapter will dive into the three main systems for an off-grid water supply: groundwater wells, natural sources, and rainwater harvesting. When you understand the pros, cons, and specific requirements for each of these, you can make the best decision that fits your needs and your surroundings.

Groundwater Wells

Groundwater wells give you a highly reliable source of water as they draw directly from underground aquifers, and this is already a go-to method for rural and remote homes. Drilling a well involves penetrating the earth and installing a pump that will bring up water from the aquifer, but it doesn't stop there. Testing the water is crucial to making sure that it is suitable for consumption. Understanding these systems is crucial, as you will have to perform regular maintenance, which includes pump checks, sediment removal, and disinfection.

This maintenance is vital to ensuring the health of the well and the safety of the water:

Pros

- **Reliable:** You get a consistent supply of water all year, which means you won't have to rely on good weather conditions.

- **Quality:** Groundwater is filtered naturally and is mineral-rich.

- **Secure:** Because the aquifers are so far underground, the water is less prone to contamination, especially when compared to surface sources.

- **Independent:** You will be self-reliant on your water needs, which minimizes any dependence on an external source.

Cons

- **High initial cost:** Drilling and installation of a well can be expensive, and this will need to be done professionally.

- **Permitting:** Local regulations and permits will likely be required. Again, always make sure before drilling.

- **Maintenance:** Regular maintenance is vital to make sure your well works at optimal levels and that you're getting quality water.

- **Potential scarcity:** While this method is overall reliable, there is still the possibility that the aquifer can be depleted, which means you will have to implement water conservation methods.

Natural Sources

Springs, seeps, streams, and even lakes can be sources of natural water for your homestead. You should always explore your surroundings to identify potential sources. However, you have to keep in mind that laws and regulations may be in place, like water rights and riparian zones. You will also need to develop skills in treatment and filtration, which are necessary because of the contamination risks that are present.

Pros

- **Renewable:** Natural sources are replenished through natural cycles, which can make these almost as reliable as well.

- **Cost-effective:** Utilizing these sources is much more budget-friendly than drilling a well, especially for an upstart homestead.

- **Gravity-fed systems:** You can use gravity to move water, which also saves money and maintenance time that you would have to put in with pumps.

- **Aesthetic appeal:** Natural water sources just add a layer of beauty to your property.

Cons

- **Variability in flow:** The natural flow of water can change significantly depending on the season and precipitation.

- **Water quality:** You will need to frequently test and treat your water to ensure that it is safe for use.

- **Legal restrictions:** Water rights and regulations can restrict or limit how much water you can use or if you can even have access.

- **Contamination risks:** Open sources like this are always subject to contamination from upstream sources or human activity.

Rainwater Harvesting

Most preppers learn about rainwater harvesting, but off-grid life takes that to the next level. Large-scale rainwater harvesting is the process of collecting and storing rainwater for your various needs. You will utilize catchment systems, gutters, and storage tanks. However, it should be noted that you can't rely solely on this method. Rainwater harvesting can be a significant source of supplementation for your other water sources, especially if you choose an area with consistent rainfall.

Pros

- **Sustainable:** You're using a readily available and renewable resource.

- **Cost-effective:** While you'd be using this to supplement your other sources of water, this can reduce the use of your other sources.

- **Reduced water demand:** While your cost might not be affected (i.e., wells using solar panels to operate), you can ease the use of your groundwater resources, avoiding any sort of scarcity.

- **Emergency backup:** Rainwater harvesting can give you a solid backup source during periods of drought or if you're having issues with your well.

Cons

- **Initial investment:** Setting up your catchment system and storage tanks requires a potentially hefty upfront cost.

- **Maintenance:** You will have to regularly clean and maintain your gutters and storage tanks to avoid contamination.

- **Climate-dependent:** Of course, you'll have to rely on significant rainfall, which can make large systems ineffective for dryer regions.

- **Water quality:** If you are using this water for potability, then you will have to filter it and treat it.

***

The right water source for your homestead depends on several factors. You'll need to consider your budget, location, and water needs. Carefully weigh out the pros and cons of each system, and consult with your community or a professional if you need to. For example, while a well is a pricey option, it's the often-chosen method for off-grid homesteads.

3.2 Water Storage Reservoirs

For anyone venturing off-grid (even if you have a well), reliable water storage becomes a huge aspect of self-sufficiency. You could have issues with your main water source, and you can't be without this important element, so now you are tasked with sourcing, storing, and purifying your water. In this section, we will explore water storage solutions when you're off-grid, both above-ground and below-ground systems. This will let you choose a system that best suits your needs and available resources.

Above-Ground Tanks

Above-ground storage is a very straightforward approach to storage. These tanks are made from various materials, and each will present its own advantages and considerations for you.

- **Plastic:** It's lightweight, affordable, and readily available. Plastic tanks are good for smaller needs. The biggest drawback is that they might be susceptible to UV degradation and leaching chemicals into the water, especially if you leave them in direct sunlight for prolonged periods. You should look for food-grade plastic and think about taking UV protection measures for your tanks.

- **Fiberglass:** This is a bit sturdier than plastic. Fiberglass is more resistant to UV and will offer a longer lifespan. The drawback here is that they can be heavier and more expensive than plastic.

- **Metal:** You'll typically find these made from galvanized steel or stainless steel, and they are extremely durable. However, they are your priciest option, and they can rust if not properly treated.

After your material considerations, you should think about the following factors:

- **Size:** Think about your daily water needs and factor in potential emergencies. Larger tanks are going to take up more space and have a higher price tag, which makes them inefficient if you're investing in a well.

- **Location:** You will need a level surface away from contaminants and direct sunlight. You will also need to make sure that you can access them easily for filling, cleaning, and other maintenance.

- **UV Protection:** When choosing plastic or a non-UV-resistant material, you'll need to build shade structures or use opaque covers to minimize exposure, which can cause quality issues.

Below-Ground Tanks

If you have larger storage needs or have specific aesthetic preferences, below-ground tanks have several options for you.

- **Cisterns:** This is a traditional, masonry-lined structure that offers natural cooling along with protection from sunlight, which can improve water quality while reducing evaporation. The issues here will be the excavation costs and potential seepage challenges.

- **Concrete tanks:** These are custom-built tanks, and they will give you unmatched durability, and you can store a lot of water. However, construction costs are steep, and like a well, you will need professional expertise.

Below-ground tanks are great, but you have to keep in mind that installation is going to be something that needs to be done right to ensure weather-tightness and structural integrity. These are also going to need regular inspections and cleaning to maintain the quality of the water.

Creative Solutions

If you don't need a large storage option, then remember to look outside of the conventional option.

- **Barrels:** Food-grade plastic barrels that have been properly cleaned and sanitized can hold water for short-term storage. As long as you ensure the lids are secure, they're protected from the sun, and you rotate the water supply, these are great.

- **IBC totes:** These are industrial-strength containers that have larger capacity and durability. Just make sure that you're using food-grade varieties and that you're properly cleaning and maintaining them.

- **Drums:** Upcycled metal drums, after they've been cleaned and treated, are sturdy storage options. However, you need to ensure they are properly sealed and protected from rust.

So, as you've seen in your smaller solutions, you need to choose food-grade materials. These containers also need to be cleaned and sanitized regularly, which means you'll also rotate your water supply. During your maintenance, you need to make sure that none of your containers have leaks or other damage that could jeopardize the supply.

3.3 Safeguarding Your Lifeblood

This is the area where preppers and homesteaders are a lot alike. The only thing that sets the two apart is that homesteaders are doing this on a grander scale. Clean, healthy water is essential for your homestead, which means you have to think about more than just turning on the tap. While that might be fine for your well system, you have to think about your storage. For this section, we will dive into the essential practices of filtration, purification, and rotation. When you use these, you will ensure your continuous access to quality water.

Filtration Basics

We start with filtration, which, as you know, is the first line of defense against unwanted particles. There are many methods out there, and the one you choose is going to depend on your water source and desired outcome.

- **Gravity filters:** This is the simplest method, but they are effective. They rely on gravity to pull water through layers of sand, gravel, and activated charcoal. This system will trap sediment, debris, and some harmful chemicals. Gravity filters are ideal for personal use and emergencies, and the biggest plus is that they are easy to maintain and don't need power to operate.

- **Sediment filters:** These filters target large particles like sand, rust, and silt. You'll often use these as a pre-filter and rarely on their own. These are ideal if you're getting water from natural sources.

- **Membrane filters:** By using microscopic membranes, these filters are powerful barriers against bacteria, viruses, and even protozoa. You can find these in various forms, with reverse osmosis being one of the most powerful you can find.

Again, you have to consider your needs. Do you need something portable? What's your desired level of purification? What are the maintenance requirements for these systems? If you are unsure, you can always contact a water treatment specialist, and they can help point you in the right direction.

Purification Techniques

While filtration can take out physical impurities, purification will tackle contaminants like bacteria and viruses. This is another step that you cannot miss when ensuring that your water is safe for consumption. The following three methods are needed when using a system other than a well. (Byrd, 2023b)

- **Ultraviolet (UV):** This method will utilize UV light to destroy harmful microorganisms without affecting the taste or chemical composition of your water. This method is effective, requires very little maintenance, and works great when used with a quality filtration system.

- **Chlorination:** This is a long-standing method, and while it can effectively kill bacteria, it will leave a slight taste in the water. It can also leave behind other contaminants. This is a good backup option for your other systems or if you're treating large volumes of water.

- **Boiling:** This is the time-tested method. Boiling water for at least a minute can eliminate most pathogens. However, boiling will not remove chemicals, and it isn't a long-term solution, nor is it a solution that's suitable for large volumes of water.

One thing to keep in mind about your filtration and purification methods is that you will still need to regularly test your water quality. It will be a bit pricey, but you should have these tests done by certified labs. They will be able to identify any potential issues, and they will ensure that you're using the most effective methods for your homestead.

Water Rotation and the First-In, First-Out Rule

After you handle your filtering and purifying needs, you have to think about storage. Just like any of your perishable items, water will degrade over time. To prevent this from happening, you will have to adopt the first-in, first-out (FIFO) rule. (Van De Walle, 2022) You'll use the oldest water first, which ensures that you have a constant flow of fresh, treated water. The simplest way you can do this is by clearly labeling your containers with the fill dates and rotating them regularly.

Cleaning is also key. When you empty a container, you need to clean it and disinfect it to prevent the growth of bacteria. When you understand and implement these practices (filtration, purification, and rotation), you'll be on your way to ensuring that your water supply is always safe.

3.4 DIY Projects for 1000-Day Water Solutions

Our first DIY section will introduce three projects that are tailored to different needs and skill levels, but all three are geared toward giving you a 1,000-day water solution. Remember to always be safe when working on any project. Wear appropriate gear and thoroughly research each project before starting.

Project 1: Rooftop Rainwater Harvesting System

This is a budget-friendly project that harnesses the power of falling rain. This is a simple, yet effective way to get a significant amount of water storage. It's a project that's great for beginners and doesn't come with a high price tag.

Materials Needed

- Large barrel (food-grade plastic or galvanized steel) with a lid that provides a proper seal.

- Gutter downspout diverter

- Mesh screen

- Spigot

- Drill

- Sealant

- Food-safe tubing

Instructions

1. **Install the diverter:** You'll start by attaching the downspout diverter to your gutter, which will direct rainwater into the barrel.

2. **Prepare the barrel:** Drill a hole near the bottom of the barrel to make room for the spigot. You'll drill another hole near the top for overflow. You'll install the spigot and seal the connections with the sealant.

3. **Install the filter:** Cut the mesh screen that's bigger than the barrel opening and secure it over the top of the barrel with either bungee cords or another material. This screen will keep out large debris.

4. **Maintenance:** To maintain your barrel, you'll regularly clean the mesh screen and your gutters. Drain and sanitize the barrel every 3–4 months with your chosen sanitation method. Keep in mind that this water is not ready to drink. You'll still need to filter and purify it if you're going to use it for human consumption.

Project 2: Elevated Gravity-Fed Storage System

This is for large homesteads, multi-story homes, or large open spaces. The benefit of this project is having consistent water pressure without having to rely on pumps.

Materials

- Large storage tank with lid and stand (This should be food-grade plastic or galvanized steel.)

- Food-safe tubing and connectors

- Filtration system (you can use any system you want, but a sand filter or carbon filter is typically a top choice.)

- Sink faucet (this is completely optional, but a great addition.)

- Pipe clamps and supports

- Sealant

Instructions

1. **Place the tank:** You'll find a sturdy, elevated location like your roof or a platform to ensure that there will be gravity-fed pressure. Secure your tank firmly using supports and brackets.

2. **Connect the pipes:** If your tank doesn't already have a hole in the lower part of it, you'll need to drill one out. From this hole, you'll insert food-safe tubing or pipe into the tank and seal it, then run this to your desired usage points. You can also install a filtration system along the line for purified water.

3. **(Optional) Install the faucet:** You can choose to mount the faucet directly on the tank, or you can install it at the end of the pipe for easier access.

4. **Maintenance:** You'll need to add the water yourself, so make sure that the tank is in a place that is easy to reach. Top off your water levels as needed. Make sure to clean the filters and sanitize your tank regularly, as you would all of your storage tanks.

Project 3: Multi-Stage Water Treatment System

This is going to be for the advanced DIY homesteader if you're looking for the ultimate water purification system. This multi-stage system gives you comprehensive treatment, even for sources that are potentially contaminated (i.e., lakes where the water is stagnant).

Materials

- Sediment filter

- Carbon filter

- Reverse osmosis (RO) unit

- Ultraviolet (UV) sterilizer

- Food-safe storage tank

- Pumps (optional)

- Plumbing connectors and tubing

Instructions

1. **Assemble your stages:** You will need to connect your filters, and they have to follow this order: sediment filter, carbon filter, RO unit, UV sterilizer. Each of these will remove different contaminants. (Byrd, 2023a) The water needs to start with the sediment filter and end with UV sterilization.

2. **Connect to water source and storage:** Connect the system to your water source and a clean storage tank. You can use pumps to maintain pressure. This is like a home purification system, and if you're more advanced, you can make this system available for your home.

3. **Maintenance:** You'll need to replace the filters regularly based on the manufacturer's recommendations. If you're using pumps, you'll regularly clean and maintain them, too. You'll also need to inspect the UV unit to ensure that it's working properly. As with other water storage, you will need to test the water quality periodically.

Keep in mind that these are simplified outlines, and while there are some who can pick this up with simplified instructions, you may need to do further research before attempting these projects.

Bonus Tips for Off-Grid Water Success

To wrap up our chapter on the lifeblood of your homestead, there are some other things to keep in mind before getting out there.

- **Conserve water:** Even if you have a well dug or have constant access to a natural source, you should always implement water-saving practices. Use low-flow fixtures inside the home, and use harvested rainwater for the bulk of your non-potable uses.

- **Monitor usage:** This doesn't mean accounting for every drop that is used, but trying to keep track of water consumption to see if there are any areas for improvement. Maybe there are areas where you could use less water or pull from one of your other sources.

- **Have backup plans:** Speaking of other sources, you should never rely on just one water source. Again, wells or natural sources are great, but if something major were to happen to your filters or your system, you still need access to clean water.

- **Share and learn:** Use your community of other homesteaders to get useful insights or helpful hints about how to make your system more efficient.

***

The story of Cape Town and Day Zero should have served as a stark reminder that our access to water is precious, and it's something that you have to consider greatly when moving to an off-grid homestead. Again, many homesteaders will invest in a well because, for long-term survival, they aren't going to take any chances. Those who live and have access to natural resources will instead invest in larger storage tanks. However, whatever system you choose, you should never rely on just one.

Having backup solutions is what DIY homesteading is all about. Maybe your backup is a storage system of harvested rainwater. No matter the solution, your other big consideration and skill-building opportunity is learning how to filter, purify, and rotate large amounts of water. Remember, you are farther out from other people, which means you can't risk yourself or someone else drinking contaminated water. Finally, remember to expand your learning. There are many more tools and projects out there regarding water, so take the opportunity to soak in as much knowledge as you can.

But now, we have to step beyond water needs and look to the next most important need for the homestead: food. As a homesteader, you will have a prepper pantry, but you're going to need to learn how to hunt, fish, and forage to keep the homestead fed.

Chapter 4:

Beyond the Prepper Pantry

Note: If you are unfamiliar with the prepper pantry, I would suggest brushing up on this concept before continuing with this chapter.

A prepper pantry is something that I would recommend that every household should have. It's a haven when faced with an unforeseen emergency or food shortage. These stocked shelves will give you a sense of security in the leanest of times. One of the realities of off-grid living goes beyond just a few weeks. There is always the potential that you're going to run into months-long shortages, and these months will see your pantry become a safety net instead of a stepping stone. If you are trying to thrive in your new landscape, you have to step into the hunter-gatherer mindset, where you will take your sustenance straight from the land.

As a homesteader, you'll be foraging for wild edibles, stealthily tracking prey, or casting a line for the catch of the day. However, these aren't things that you will just master overnight. These skills require you to adopt the homesteader mindset and to get a deep understanding of the surrounding environment.

Remember the initial steps into food storage? You always start with a few days' worth of food, then you gradually expand to a few weeks. Along the way, you're learning about shelf life, rotation, and efficient storage methods. Picture this initial journey extended to a few months, even years. Your prep pantry is still going to play a role, but you're going to go much further on this journey.

In this chapter, you will make your transition from a consumer to a very active participant in the food chain. I will go over the challenges and rewards of hunting, fishing, and foraging. You will also learn how to navigate the legal landscape and understand sustainable practices.

A word of caution, though, before we move on. Don't wait until you make your move off-grid to hone these skills. That will present a huge challenge to you while you're trying to establish your homestead. This is a situation where practice makes perfect, especially if the learning curve is steep. Start small with local workshops, and then move on to connecting with experienced hunters, fishermen, and foragers. The more you can learn and practice, the smoother things will be when you do make your move onto the homestead.

4.1 The Basics of Hunting

When envisioning an off-grid lifestyle, the image of hunting likely springs to mind. It's part of the hunter-gatherer mentality, and it's always been an image of survival, living off the land, and providing for oneself and their family. It's a primal image, but it is something that you will partake in. However, before you step out with your hunting gear, you have to master the basics.

This is essential because hunting isn't just about survival; it's about the respect we have for nature, upholding ethical hunting practices, and understanding the legal landscape. So, let's dive into the fundamentals, the challenges, and why this skill is something that you will want to be a cornerstone in your off-grid transition.

The Challenges of Hunting

The following are some of the main challenges you'll face when taking on the role of off-grid hunter:

- **Know the game and its habitat:** Imagine that you head out and track an animal for hours. However, right before you make the kill, you discover that it's protected. This can be disheartening, but that's why understanding the game in your area is extremely important. Learn what species inhabit the area, their identifiable traits, which of them are safe to eat, and which of them are safe to hunt. You can use field guides or consult wildlife biologists. You should also spend time out there observing animal behavior in your area. Knowledge is going to be your strongest ally, not just instinct.

- **Tracking and stalking:** It's an exhilarating feeling to be chasing an animal, but you can't forget to be responsible. Getting caught up in the chase could lead to you venturing too far from the homestead, and without proper navigation skills, this could be trouble. You need to learn how to track efficiently while respecting animal behavior. This will prevent unnecessary stress from being put on them. You also need to understand the land around your homestead. The terrain, water sources, and potential hazards are all great things to know. Also, keep in mind that tracking and stalking are skills that require a lot of patience, planning, and an understanding of the animal's habits. Never underestimate the challenge, and hone these skills before making your move off-grid.

- **Gearing up:** Not all arrows are made equal, and neither are rifles. Not all traps are suitable for every animal. There is such a thing as "too much" when hunting, so research the appropriate equipment for the animal you're hunting *and* the surrounding land. Consider things like weapon legality, the effectiveness of your ammunition, and ethical practices. Traps are something that requires specific knowledge and practice before being used to remain humane and avoid any unintended consequences. While you want your hunting tools to be efficient, you also have to be responsible and, above all, ethical.

- **Being ethical:** It's a way to provide food and other materials, but it's still taking a life. It's a serious act; therefore, your actions should all stick to ethical principles. You need to prioritize clean, humane kills that prevent any suffering. This is why you need practice before moving off-grid; things like shot placement and appropriate ammunition are going to help you stick to those principles.

- **Butchering and processing:** Hunts don't just end with a successful kill. The most crucial part of this endeavor is the timely processing of the animal. You will need to learn proper butchering techniques to get the most yield while producing the least amount of waste. You'll need to consider using all parts of an animal—organs for stews, hides for leather work, and bones for broth or tools. This is an area where your preservation skills (drying, smoking, and canning) will come into play to ensure you can use everything you harvested.

Rules and Regulations

Knowing the legal landscape is a must-have knowledge. Even in this disconnected lifestyle, there are still laws that you have to obey. These things are in place to protect the animal population and to ensure that you remain an ethical hunter. Let's get into some of those regulations that you will need to keep in mind (Rinella, 2018).

- **Licenses and permits:** Unless we really hit a societal collapse, hunting and trapping are practices that require licenses and permits for your region. These are the things that you will need to do to go on guided hunts. You will also need to know the renewal dates and any applicable fees.

- **Bag limits and seasons:** Each species in your area will have a designated hunting season, and you will be limited in how many animals you can kill per hunt or season. This ensures sustainable populations, and ignoring these regulations will lead to hefty fines and legal trouble, including your ability to hunt in the future.

- **Conservation matters:** You will maintain a healthy ecosystem by using responsible hunting practices. Take time to become familiar with conservation laws and follow ethical harvesting practices.

- **Respect your property lines:** Know your property lines and know who owns the land surrounding you. Never hunt beyond your property lines without getting permission from those landowners. Breaking this rule can lead to legal repercussions, and it can cause friction within your off-grid community.

Why Hunting Is Necessary

Some new off-gridders might question why hunting is a necessary skill to learn. Despite the challenges, hunting offers homesteaders valuable benefits.

- **Sustainable source of protein and fat:** Wild game is nutrient-rich with proteins and fats, which are key in long-term survival situations. It's a great supplement to your diet, especially in times when plant options are limited.

- **More than just food:** You can use animal materials for various things. Bones can be forged into tools, hides can be tanned for warmth, and sinew can be used for sewing and various crafting purposes.

- **Skill development:** Hunting will build up essential survival skills, including tracking, marksmanship, and general resourcefulness. These skills will transcend just the need for food.

4.2 The Basics of Fishing

If you are fortunate enough to have your homestead near water, fishing is an amazing alternative (or supplement) to hunting, especially if there is a healthy population of fish in that water. While hunting tasks you with being stealthy and having good tracking skills, fishing will present its own challenges, but often with a higher success rate. Even beginners find faster success with fishing. Much like hunting, we will dive into the fundamentals to ensure you enjoy a safe, sustainable harvest.

The Challenges of Fishing

While there is a higher success rate with fishing, there are still unique challenges presented when using this method.

- **Location and technique:** The key to your success is understanding what fish are present and the environment. Much like with hunting, you'll need to research what fish species are in the water and their preferred habitats (rocks, weeds, etc.). You'll also need to learn the best techniques (bait, lures, fly fishing) for catching them. You can learn a lot of this if you practice before making your move off-grid.

- **Baiting and casting:** Baiting your hook and casting your line is an art that takes a lot of time and practice to master. Spend some time learning different knots that will make hooks secure but easy to remove. You can also practice your casting accuracy in other safe areas before trying it out on the water. You want to do this because fishing is more about finesse than brute force, especially if you want to attract a good haul.

- **Identification and preparation:** An overlooked aspect of fishing is that knowing your catch is crucial. Look for a fish identification guide for your region and learn how to not only identify protected species but also which fish are edible. Once you know what types of fish are available to you, you will want to be familiar with how to safely handle them and how to properly clean them to ensure no one is getting sick.

Rules and Regulations

Rules and regulations also apply in the world of fishing, and while the following are things to keep in mind, you should always check with your local game office to know all the rules for your region. (Burnley, 2021)

- **Licenses and permits:** You should always get the proper fishing licenses and permits for your location and body of water. Failure to do this will not only result in fines but can severely damage the ecosystem.

- **Catch limits and seasons:** As you learned with hunting, these regulations are in place for good reason. They ensure that the fish population is sustainable. Always respect catch limits and stick to permitted seasons. When you only take what you need, you are allowing the waters to remain teeming with life.

- **Catch and release:** Unless you are in a deep survival situation, you will only need to catch and keep enough for food. However, catch-and-release practices allow you to keep your skills sharp while allowing the fish to grow, reproduce, and maintain a steady population.

- **Navigational regulations:** When fishing from a boat, you should be familiar with these regulations and learn what areas of the water are designated boating areas. Following these rules will keep you and others safe on the water.

Why Fishing Is Necessary

It takes time to get a bite, which means that fishing is a skill that requires a lot of patience. With everything else there is to do on the homestead, why would one want to use this practice?

- **Nutrition:** Fish pack a wealth of protein, essential nutrients, and omega-3 fatty acids—all necessary components for a healthy diet. A few sporadic fishing excursions can contribute so much to the nutritional needs of your home.

- **Sustainability:** When compared to hunting, fishing is a much quieter, low-impact way to get the foods that you need. The time it takes to have new fish in the water is significantly shorter than the large animals you would hunt, and you have the catch-and-release option, which contributes to aquatic ecosystems.

- **Developing skills:** Fishing will hone valuable skills like patience, resourcefulness, hand-eye coordination, and even problem-solving. You're also going to have a deeper connection to nature when you learn about your local ecosystem.

4.3 Basics of Foraging

Again, off-grid living brings up the idea of the traditional hunter-gatherer role, and so we have to round this off with the gatherer aspect. Venturing into the world of foraging, though, is going to be much more than having a backpack and a sense of adventure. It, too, poses its own set of challenges, rules, and regulations. It's a demanding practice, but this section will get you ready to tackle it with ease.

The Challenges of Foraging

Foraging doesn't present the same level of challenges that hunting and fishing do. However, the most significant hurdle is being able to confidently distinguish edible plants and mushrooms from their harmful lookalikes. Unlike your conventional way of buying produce, nature doesn't give you neat labels for everything. Instead, you get an astonishing array of flora, with many plants looking similar to other ones. There is very little room for error, so mastering accurate identification is a must. To have a better chance out there, you will need to invest in the following.:

- **Field guides and resources:** Invest in reputable guides that were written for your specific region. These guides need to have detailed illustrations and descriptions. You could also use the USDA "plants" database to get verification.

- **Expert guidance:** I would strongly suggest seeking help from experienced foragers or looking for a foraging group in your area. This will give you valuable insights along with hands-on experience, which you will need before moving off-grid.

- **Start small and slow:** This is a challenge because a lot of people just want to get out there and forage for everything. But you should start with the easily identifiable plants, gradually building your knowledge as you gain more confidence out there. Some of the best advice that I can give you when you're out there is to never eat anything that you're not sure about, and if you're in doubt, throw it out.

Foraging Responsibly

This skill isn't just about taking from the land; it's about living in the balance of the ecosystem. You also want to ensure that this bounty continues to be there, so the following list will list ways that you can do this.

- **Mind the season:** You should respect the natural growing seasons and cycles of your plants. You should only harvest when plants have reached their peak, which ensures there will be continued growth. Those peaks are things you will learn from your guide and local groups.

- **Leave no trace:** Avoid stripping plants bare. By taking only what you need, you allow them to replenish while supporting other wildlife in the area.

- **Use sustainable techniques:** You can use harvesting methods like selective picking, but you should always be mindful not to damage the root systems of plants.

- **Respect protected plants:** Make sure you know if there are any endangered or protected plant species in your region. These are plants that you will avoid, as this contributes to your local conservation efforts.

Rules and Regulations

Of course, there is a legal framework around foraging, too. Remember to do research and become familiar with this to ensure that you're foraging the right way. (Ana, 2022)

- **Public land:** Foraging on public lands might come with heavy restrictions, or you may need to file for a permit. Do your research on your local region and make sure you have the necessary permission before going on public land.

- **Private property:** You should treat other lands like you treat your own. If the best foraging areas are on someone else's property, you should have permission from them to avoid trespassing.

- **Protected areas:** When doing your research, you will be able to designate reserves, sanctuaries, and areas where foraging is strictly prohibited. These places are typically in place to protect endangered plants and fragile ecosystems.

Why Foraging Is Necessary

Despite the challenges that you will run into while foraging, there are a lot of positives that you can take away from it.

- **Diet diversity:** Wild plants are dense with vitamins, minerals, and fiber. There is also little processing, and you can eat these fresh, which is enriching for your diet. Not to mention that they are packed with flavor.

- **Enhanced self-sufficiency:** Foraging lets you supplement your other food sources, which gives you a greater sense of security and independence, especially when hunting and fishing run into dry spells.

- **A deeper connection to nature:** The act of foraging allows you to connect more to the world around you, which offers new insights and allows you to find a better balance.

4.4 DIY Projects for 1000 Days of Meeting Food Needs

The first 1,000 days of using hunting, fishing, and foraging techniques will be an exciting adventure, and you are going to need to call on your resourcefulness to do it. These DIY projects were designed to be used immediately, which gives you a chance to build your confidence while procuring your own food. Of course, remember to prioritize safety and handle your tools and materials with care. Another thing to keep in mind is to double-check that there are no regulations against these projects.

Project 1: The Minnow Trap

This trap is one that any beginner can use, and it only requires a few simple materials that you likely already have lying around your home.

Materials

- Plastic bottle (1 liter is fine; 2-liter bottles are better)

- Scissors or a knife

- Wire or string

- Mesh fabric (optional but will increase durability)

- Bait (your preferred choice will do)

Instructions

1. Start by cutting the bottle. You'll cut the top third of the bottle off, which is going to make your funnel-shaped entrance.

2. Cut slits in the bottom of the remaining section. This is going to allow water to flow through the trap.

3. Now, you're going to secure your funnel, and you'll do this by inverting the section you cut off and inserting it into the "bottom" part of the bottle. Secure this with your wire or string. This is how you'll form the continuous trap.

4. It's time to bait and deploy, so thread baits through the funnel and secure them inside. You'll tie the trap to a stick or a weight and partially submerge it in the flowing water.

5. Check your traps regularly and gently collect the minnows. You can then rinse and store them, using them later as bait for your bigger catches.

Maintenance

Maintenance on the minnow trap is fairly easy. First, raise the trap after each submersion to stop odor and buildup. You'll also replace the bait to ensure that it's always effective. If you opt to use mesh fabric, you need to make sure that it is clean and free of tears.

Project 2: Natural Fishing Line

While you can hope that you have plenty of durable fishing lines in storage, you have to think about sustainability. This brings up the need to craft your own line from natural materials. This project is going to take much more patience and a lot of practice, but once you have made your first line, you won't want to go back to store-bought lines.

Materials

- Plant fibers (yucca, milkweed, or any plants in your region that have durable fibers)

- Rolling pin (a smooth stone will also work)

- Water

Instructions

1. Gather your plant fibers. You'll want to make sure these fibers are strong and flexible. Remove any weak fibers, leaves, and other debris.

2. Using your rolling pin (or stone), you will roll the fibers back and forth. This is going to break them down, and you'll begin to extract the stronger inner strands.

3. When you have extracted the fibers, twist the individual strands together. You'll start thin, but then you'll gradually add more to make the line thicker. Remember to test the strength of the line regularly.

4. Start lengthening the line. Continue twisting and adding fibers until you get the length that you want. If you need to, you can tie multiple lines together.

5. Take your line and soak it in water. This will improve the line's flexibility, and after it's soaked, dry it thoroughly.

Maintenance

You'll want to store the line in a cool, dry place with no exposure to direct sunlight. Regularly check the line for damage and use this time to replace any sections that need it. Also, don't stop with one line! Try out different plant materials and techniques that work for you.

Project 3: The Tilt Trap

This project is going to be one that you can use when you're a bit more advanced in your journey, but despite the difficulty, the rewards are immense. A well-built trap can effectively capture small game while remaining ethical. Keep in mind that you will need to check local regulations, as you might need to get permits before setting traps.

Materials

- Sturdy logs or branches

- Tripwire or string (this will be your trigger mechanism)

- Bait (fruit, small scraps of food)

- Something to use as a platform or a base

Instructions

1. Start by making an A-frame with the logs or branches. Your frame needs to be stable and secure.

2. You will choose a trigger mechanism that will release when weight is applied.

3. Build a platform. This is going to be what holds the bait and triggers the mechanism.

4. All that's left to do is secure the bait to the platform and camouflage the trap with natural materials that will help it blend into the environment.

5. You'll regularly check the trap and release any non-target animals.

Maintenance

You will inspect the trap during your checks to make sure there is no damage, and you'll repair it as necessary. You'll also make sure that the trigger mechanism maintains its sensitivity. The most important thing to do, though, is to clean the trap and change the bait regularly to avoid unwanted pests.

***

Acquiring food is one of the crucial skill sets needed when you move off-grid, and in this chapter, we have covered the fundamentals of three of those methods. Each method is going to come with its own set of challenges, but they all have their rewards. As long as you respect the environment and your local regulations, you will see a lot of success. Let's look again briefly at those before we move on.

Hunting provides your homestead with protein and other valuable resources, but you have to use ethical practices. Do as much as you can to learn about local game, proper tracking practices, and responsible harvesting techniques. Make sure you have the proper licenses and permits, too.

Fishing is quieter and more accessible while still providing you with the nutrients you need. That doesn't mean that there is no challenge. You will need to have a handle on baiting techniques, casting, and identification of the fish in your region. Again, licenses and permits will be needed, and you should always respect catch limits.

Foraging will help you expand your diet, and it will give you a lot of vitamins and minerals that you won't get from hunting and fishing. However, this might be the most challenging method because accurate identification is crucial. This is why you will need to start slow, invest in local guides, and seek local experts. Be mindful of property lines and secure permits if you have access to public foraging lands.

These are great starting points to supplement your prep pantry, but now we need to expand beyond that. In the next chapter, we will explore the world of growing your own food.

Chapter 5:

Making a Sustainable Garden

Hurricane Maria wreaked havoc on Puerto Rico in September 2017. Maria left behind leveled homes, downed powerlines, and just a landscape of widespread devastation. This also meant that crops were flattened, and grocery stores were either destroyed or inaccessible. Many were left hungry, and they faced the huge challenge of securing food. With so much despair, it seemed too daunting, but a sign of resilience emerged in the form of community gardens. (Holpuch, 2018)

Many people worked to cultivate these lush patches, which ultimately served as Puerto Ricans' lifelines. Residents even rallied around community gardens that had been established well before the devastating hurricane came ashore. These gardens were established for a variety of reasons, but they would prove to be invaluable in the wake of Maria. Their determination is now fueled by feeding not just themselves but also their neighbors. Together, they would share seeds, tools, and other knowledge, which allowed more people to turn the devastated land into a source of fresh food.

More stories like this sprouted up across the island. Gardens in Caguas provided needed food for families that were struggling to get by. Community healing and the process of rebuilding started in the gardens in other towns, too. So, the gardens for the people of Puerto Rico were more than just plots of land with plants; they became symbols of unity and resilience. They also stood as a reminder of the power of collective action.

The impact of these gardens, though, extended far beyond the immediate relief they provided. They became reminders that people can grow gardens large enough to feed a community while being off-grid. Where their infrastructure was crippled, these gardens became a sustainable solution, empowering communities to become self-sufficient. Residents began growing their own food, which cut down on their reliance on any external means.

The lessons learned in Puerto Rico after that disaster can hold valuable insights for anyone making the switch to an off-grid lifestyle. The following things should be kept in mind as we navigate through this chapter:

- **The key to success in the community is collaboration** and sharing knowledge, which are very beneficial. This is why you should connect with others, even if they aren't in your area. You can learn from their experiences, and you can build a support system.

- **Start small and adapt:** You shouldn't approach this thinking that your first couple of harvests will provide all the food you need. Don't overwhelm yourself, and start with a manageable plot on your property. Afterward, experiment with different crops and learn from your successes and failures.

- **Be resourceful:** Use things like harvested rainwater, compost, and other natural techniques that will cut down on your waste and promote healthier soil and plant growth.

- **Be diverse:** Learn how to plant a variety of crops and know optimal growing seasons to allow for an extended harvest period. This will also help your garden be flexible in various conditions.

In this chapter, you will learn how to test and maintain healthy soil, how to plan your off-grid garden, how to care for your plants, and how to compost to add nutrients back to your soil. This chapter is also going to stray from the formula, as the DIY projects are incorporated into the chapter. With that said, let's start growing!

5.1 Understanding Your Land

Before you sow your first seeds or raid your first seedling, your garden journey on your homestead starts with understanding the very ground you walk on. In this section, we are going to cover two crucial aspects of crop planting and gardening: hardiness zones and soil conditions. Understanding these will allow you to make better choices that lead to more plentiful harvests.

Hardiness Zones

Imagine planting watermelon seeds from where you lived in the cooler climate of your new homestead. Your results would be lackluster, to say the least. This is why you need the Plant Hardiness Zone Map from the USDA (Boeckmann, 2024) (there are maps for other parts of the world, too). This coded map will predict what plants will grow in your region based on the region's average minimum winter temperatures. You'll be able to find this map easily online or at your local garden center.

Each zone on the US map is represented by a number (3–11). Lower numbers represent your colder climates, and higher numbers represent the warmer zones. So, a gardener in Zone 3 will know that more tropical plants will not survive in their zone, and a gardener in a more tropical setting will have difficulty with leafy greens. There are some climates, though, that provide a "sweet spot" where the non-growing season is very short. (Boeckmann, 2024)

You should keep in mind that this map is just a general guide. You can use microclimates within your property to open up more room to experiment.

Microclimates

Don't fret too much if your hardiness zone limits your growth because there is power in microclimates. These localized areas on your property can have slightly different temperatures, light exposure levels, and moisture conditions—all ready for you to utilize them.

South-facing slopes get more sunlight, making them warmer, which means you'll have a microclimate that's ideal for heat-loving plants. Then you take on your north-facing sides; these areas are cooler, which is great for delicate greens. It will take some practice, but when you get better at using microclimates, you will push the boundaries set by your hardiness zone and your garden's potential (Brotak, 2020).

The Power of Healthy Soil

The foundation of your garden's success is going to come from the soil. Healthy soil is so densely packed with beneficial microorganisms, which are going to provide essential nutrients to your plants, and they will anchor your plants. There are different types, and each is going to have its own unique characteristics and will need its own specific approaches. Let's look at those types and their ideal crops.

- **Clay soil:** This soil is dense and nutrient-rich, but it compacts easily, which causes drainage issues. You can fix this with organic matter like compost, and you can improve the drainage by using raised beds. The best crops for this soil are root vegetables, leafy greens, and some herbs.

- **Sandy soil:** This soil drains well, but it is scarce in the nutrient department. You can amend that with organic matter and compost it regularly. This soil is great for drought-tolerant crops like peppers, melons, and certain herbs.

- **Loam soil:** This is the most ideal soil that you can have on your homestead. It has a well-balanced texture, good drainage, and is extremely fertile. You can grow a variety of fruits, vegetables, and herbs in this soil.

Project 1: Soil Testing and Amending for Optimal Growth

You might be considering amending your soil, but before you start adding things to it, you should know where you're starting from. This will allow you to make the right amendments. You could send your soil to a lab, which can be pricey. Or you can make your own DIY tester that will give you a rough estimate of what your soil type is, and all you need is a jar.

You'll start by filling your jar a third of the way with soil, and you'll top the rest off with water. Your next step is to shake the jar vigorously and watch the settling layers.

- **Sandy soil:** Sand will settle quickly at the bottom.

- **Clay soil:** There will be a very distinct, cloudy layer above the sand.

- **Loam soil:** You'll see more balanced layers, with some sand, silt, and clay.

Based on what your tests say, you can then amend accordingly.

- For sandy soil, add organic matter and aged manure for moisture retention and nutrients.

- For clay, add matter like compost, aged manure, or leaf waste to help drainage and aeration.

- Loam soil is great, but you still need to amend it. You can do this by regularly adding compost or aged manure. This helps maintain the balance.

Beyond the Basics

Remember that this is just a springboard for your homestead's garden. The deeper you get into this, the more you start considering other things like rainfall patterns, sunlight exposure, and prevailing winds. Understanding your land is going to be an ever-evolving process, so don't be worried when the learning curve is steep.

5.2 Planning the Off-Grid Garden

Having an off-grid is a great step in providing long-term sustainability for the homestead, especially when your eyes are focused on those first 1,000 days. But now comes the really tricky part: planning the garden! Few growers consider this, but an effective garden is so much more than throwing some seeds in the dirt and waiting for them to sprout. A carefully planned layout will set you up for bountiful harvests while keeping wasted resources to a minimum. We will use this section to familiarize you with the essentials of planning out your garden.

Zone Planning

Picture the plot of land divided into zones, and each of those will have a specific purpose. This is the concept of zone planting, which is a strategy that maximizes your usage of space while catering to the many different needs of various plants. So, how does the concept work?

- **Sunlight zones:** Divide your garden based on sunlight exposure. Sun-loving plants like tomatoes and peppers should be in the "full sun" zone, while leafy greens like lettuce and spinach should be in the zone that has some shade.

- **Water zones:** You will also group plants that have similar water requirements. Your water-dependent crops, like cucumbers, will do great being closer to your water source. More drought-tolerant plants can be placed farther away in drier areas.

- **Plant size zones:** Research the mature size of your plants. A taller plant like corn shouldn't shade something like carrots. Make sure you allocate the space accordingly and do your best to avoid overcrowding your plants. This will keep them from competing for resources, resulting in a poor yield.

Project 2: Zoning Different-Size Gardens

I would suggest carefully considering this project. Every garden is going to be different, so knowing about your zones will allow you to divide your space effectively. The following are just a few examples of what you could use on your homestead.

- **Small garden (around 100 square feet):** This size is great for three zones. You'll have a sunny zone for sun-loving plants, a partially shaded zone for your leafy greens, and the last section dedicated to herbs. If you want to maximize space, use vertical gardening techniques like hanging baskets and trellises.

- **Medium garden (around 500 square feet):** You can expand to five zones. You'll have your full sun zone, your partial shade zone, and your dedicated herb zone. But now you'll get to add a trellis-specific zone that will be for your climbing vegetables. Your fifth section will be a nitrogen-fixing legume zone (beans and peas).

- **Large garden (1000+ square feet):** Stretch your garden out to some very unique zones! Your zones now will be for different plant families—a solanaceous zone, a cucurbit zone, a brassica zone, an herb section, a zone for legumes, and you'll even have room for a fruit tree area.

Remember that these are starting points, and you don't have to follow a rigid plan. Adapt your zones to fit your needs and preferences. You can get a lot of help planting in your specific region by using online resources and even gardening apps. This will also help you develop a plan that best fits your space and sunlight conditions.

Choosing the Right Plants

So you've checked and amended your soil, and you've mapped out your zones, so now you get to choose your off-grid plants. There's going to be a lot of variety depending on where you live, but let's look at some of the key considerations.

- **Hardiness Zone:** You should keep this in mind, as this map will help you make the right choices with your plants. Always choose things that will thrive in your local climate.

- **Soil conditions:** Buy a soil test from your local garden center to test your soil's pH and nutrient levels. This is almost like what you did earlier, except you're working on the fine details. Amend your soil as needed to establish an optimal growing environment.

- **Fast-growing, high-yield:** Your first choices should be plants that mature quickly and will have abundant harvests. This will allow you to maximize food production, even when working with limited space. Cherry tomatoes, bush beans, and leafy greens will thrive if they fit your hardiness zone.

- **Companion planting:** Use this concept to the fullest of your ability! Certain plants benefit from one another because they attract beneficial insects, deter pests, or simply improve soil health. For example, if you plant marigolds near your tomato plants, they will deter insects that would harm the tomatoes.

- **Be diverse:** Don't just plant one type of plant, even if you're limited in space. You should have a healthy mix of fruits, vegetables, and herbs. This will ensure that you have a steady flow of nutrients, and it will help maintain soil health, especially when you use crop rotation.

Keep in mind that you want to start small, especially if you aren't familiar with gardening. Start with a few key crops and expand from there as you get more experience. Don't be afraid to make a few mistakes, either. The most important thing you can do is approach growing with passion, and you will achieve great success.

5.3 Successful Cultivation

There is nothing more satisfying than watching seeds grow into a bountiful harvest. Therefore, let's dive into the knowledge and techniques that you need to cultivate thriving plants and bask in the rewards of your labor. This section will cover planting methods, proper watering techniques, sunlight management, pest control, and harvesting. The goal is to have a year-round food supply strategy.

Planting Strategies

You can start your off-grid garden in many ways, but we will look specifically at three methods: direct seeding, transplanting seedlings, and starting your own seedlings.

Direct Seeding

This is the traditional method of sowing seeds directly into prepared beds of soil. This method is ideal for more hardy vegetables like carrots, radishes, and leafy greens. Pay attention to the labels on the seed packets (or look online), as this will help you sow the seeds at the recommended depth and spacing. After planting, you'll lightly cover the seeds with soil and water the area well. One thing to keep in mind with direct planting is that your soil needs to have consistent moisture for germination, which means you might need to water the soil before planting.

Transplanting Seedlings

Delicate plants, or those that need to be in a controlled environment, will be ones that you start indoors in trays or pots. Make sure you use a soil mix that's good for seed starting, then give the plants the right amount of light and warmth.

After the seedling develops its true leaves, you can start hardening the plant by exposing it to outdoor conditions. After it's hardened, you'll dig a hole larger than the root ball of your plant and place it at the same depth that it was in the pot. Keep in mind that you might need to amend the soil before planting, though. After the plant is transplanted, water thoroughly.

Starting Your Own Seedlings

This method offers a wider variety of plants, and you will save money as transplanting sometimes means buying the seedling from a garden center. Of course, you want to choose seeds that are suited to your climate and that are in season. Make sure you stick to the germination requirements for each plant type and give them the right amount of light, temperature, and moisture.

There is a bit of an initial investment in starting your own seedlings. You'll be investing in trays, grow lights, and heating mats. You'll also need to designate an area where you will start these seedlings. Some homesteaders will even have a simple greenhouse on their property just for this purpose.

Smart Watering

You know that different plants have their own watering needs, but did you know that different stages will also have distinct needs?

Seedlings need frequent, light watering to have consistent moisture. As your plants mature, you want to encourage strong root development; therefore, you'll water deeper and less frequently.

Choose to water early in the day to avoid evaporation. After the seedlings take on their first leaves, you'll switch to watering near the soil base and not on the foliage. You will also adjust your watering schedule based on the weather and soil type.

Sunlight Exposure

Sunlight is a crucial factor in your plant's growth. This is why research on your chosen plants is a must. You need to identify the light requirements of your plants and use your planting zones accordingly. If you have potted plants, practice regular rotation to make sure they have even exposure.

Natural Pest and Disease Management

Prevention is key here. This can be done by choosing disease-resistant varieties, crop rotation, and simply keeping your garden clean. You will also want to encourage beneficial insects like ladybugs and lacewings. These insects will prey on harmful pests. If you do need other pest control, aim for natural methods like insecticidal soap, neem oil, or diatomaceous earth.

Aside from prevention, being vigilant will lead to early detection. When you can catch pests and diseases as they begin, you will be able to promptly address them.

Harvesting the Bounty

Knowing when you need to harvest is another huge consideration. You should know your plants well and observe color changes, their size, and even the texture of each crop. By fully researching your plants, you will be able to harvest them for optimal flavor and quality.

Your best tools here will be very sharp ones. This will prevent damage to the plant while encouraging new growth. You can regularly harvest plants like leafy greens and herbs, as this will promote continuous production.

After your harvest, you can extend its shelf life through various techniques. You can dry things like herbs and peppers to concentrate flavor. Canning fruits and vegetables will allow them to stay fresh for months. You can blanch and freeze vegetables to ensure they keep their texture and nutrient levels. The more preservation techniques you know, the wider the variety you will have in your pantry and the longer you'll get to enjoy that harvest.

Planning a Year-Round Pantry

The best way to ensure that the homestead doesn't run short on food is to stagger your planting and harvesting times. This is going to take more research and planning, but you can choose varieties with different maturing periods. You can also choose cool-season crops like lettuce and spinach to plant in the fall for a winter harvest. This is also another reason homesteaders will use greenhouses or cold frames, so always keep those in mind if you have the space. These are just a few techniques that will let you keep a consistent supply of food.

5.4 Purposeful Wasting: The Art of Composting

Turning your kitchen scraps and yard waste into rich, dark soil is one of the most environmentally friendly acts that you can do on a homestead. At its core, composting simply mimics what happens in nature. Microorganisms like bacteria and fungi break down the organic matter, creating a nutrient-rich product that is like gold for your garden.

The key to a successful compost is to create the ideal environment for the decomposers to thrive. Even this pile of "waste" needs to have a good balance of moisture, aeration, and temperature. Think of it like building a mini-recycling plant on your homestead.

Choosing Your Method

Much like all the other work that will go into the homestead, there is not a one-size-fits-all approach to composting. You will need to factor in things like your available space, your time commitment, and how fast you want the material to decompose. The following are some common methods:

- **Pile composting:** This is the traditional method, which involves layering green (nitrogen-rich) materials like food scraps and coffee grounds with brown (carbon-rich) materials like leaves and shredded cardboard. This is a low-maintenance option, as you just need to regularly turn the pile to aerate it and speed up the process. However, it is not made for speed, as this will take anywhere from 6 to 12 months to complete. (Cal, 2021)

- **Bin composting:** This method is a lot like pile composting, except it is in a bin. This method is good for odor control, and you can protect the pile from critters. A wooden bin with holes works best, but you can always repurpose plastic containers.

- **Tumbler composting:** This method is great if you have limited space. These are sealed compartments that rotate, which speeds up the process. The only real issue is how little they hold and they can dry out faster.

Project 3: DIY Cold Composting

The best way to start composting is with this simple, low-effort option. All you need to do is clear a dedicated space somewhere on your property. After that, you'll use the pile composting method, using green and brown materials. Cover the pile with leaves or straw and just let nature run its course. This is a much slower method (12–18 months), but if you have the space, you can stagger your piles for extended use.

Hot Composting

If you want your compost done quickly, hot composting will give you what you need in 4 to 6 weeks. If you can maintain a temperature of around 55 to 75 degrees, the microorganisms will work faster and break the organic material down quickly. This method does require more attention because you will have to turn it regularly, monitor the moisture levels, and make sure that the green and brown materials are balanced. (Cal, 2021)

Project 4: DIY Hot Composting

1. Choose a well-ventilated area that gets direct sunlight and is level. If you need to amend the area, then do so before starting.

2. Using wire mesh or wooden slats (which allow for aeration), build a bin that measures 3'x3'x3'.

3. Layer your green and brown materials. I would recommend a 3:1 ratio (3 parts green, 1 part brown), but this can vary from garden to garden.

4. Turn your pile at least every other day. While you're turning, make sure that the pile is moist, not saturated. If it's dry, then add water.

5. Make sure the compost is maintaining its temperature and adjust the layers as needed.

Vermicomposting

This is an indoor method that uses composting worms to break down your materials. These worms will produce castings, which are nutrient-rich and help improve the structure and aeration of the compost. If you have limited space and need something odorless, this is the method for you.

There isn't a way to DIY this because your bin has to be specifically made for the composting worms. Most of these will also come with instructions on how to care for the pile, making this a really great option for smaller homesteads.

Tips and Tricks for Composting

Before we close off this chapter, here are just a few more tips that you should know about composting.

- Shredding or chopping your materials before layering will give you faster decomposition.

- Avoid meat, dairy, or any oily scraps. This will attract pests and make for slower decomposition.

- Sprinkle the compost around your plants or mix it into the soil to get the benefits of a nutrient boost.

- If you're having success with vermicomposting, consider expanding your worm bin to handle the population.

While there is a lot to consider with off-grid gardening, you don't have to wait until the transition to start practicing everything you've learned in this chapter. Any backyard can be zoned to grow plants, and if you're living in an urban space or somewhere without a decent yard, you can practice growing in small spaces. Any practice with seeding, growing, harvesting, and composting will set you up for more success when you do move to your homestead.

Now that we have covered the concepts of hunting, gathering, foraging, and growing, we have to turn our attention to another addition to the homestead. Our focus now goes to raising our own animals.

Chapter 6:

Thriving Off-Grid Livestock

Now, we come to a piece of off-grid life that should also be highly considered if you have the land to do so. You already know that this journey is one of self-reliance, and livestock can become an invaluable ally. Your animals are going to be so much more than barnyard residents. They are going to contribute almost as much as you will to the homestead. Imagine having hens produce fresh eggs, nutrient-dense cheese made from goat's milk, and fertilizer made by your pigs. It's almost an off-grid wonderland.

But this wonderful life isn't without hard work. Raising animals demands dedication, maybe even more than the rest of your off-grid life. You will face early mornings and a lot of physical labor. You'll have to protect your animals from nature's wrath and the predators that lurk on the land. This step in your journey requires vigilance, and it will also require financial readiness as you face start-up costs, ongoing expenses, and vet bills.

Yet, for many who bring livestock to their homesteads, the rewards far exceed the obstacles. This creates harmony between the animals, the land, and the people. In this chapter, we will cover how to choose the right off-grid companions for you, the inevitable challenges you will face, how to protect your animals and their essential needs, and the DIY projects that will help you and your animals.

6.1 Choosing Your Off-Grid Companions

Again, this is going to be a very rewarding experience for you. Animals not only provide a source of food, but they can also help tend to your land. They also add an element of companionship to the homestead. However, you can't just go and buy chickens, goats, pigs, or any other animal. This is a decision that demands your careful approach, considering the various factors before welcoming an animal onto the land.

So before bringing them home, here is a list of things you should consider:

- **Climate and land:** You should already be well acquainted with the climate and terrain of your property, from building your homestead and growing your garden. So now, you have to think about it in conjunction with animals. Some companions, like goats and sheep, are going to love colder climates, while your feathered friends, like chickens and ducks, will do better in warmer conditions. As far as your land goes, you need to consider the size. Large animals and grazing animals will need space to move, while your smaller animals can adapt to tighter living conditions.

- **Resources and experience:** You have to be completely honest about your available resources and experience level. These are living creatures, and raising them is going to take time, money, and a lot of effort. Ask yourself things like "Can I afford the initial setup costs like housing and fencing?" or "Am I prepared for the ongoing expenses of feed, vet care, and repairs to their pens?" Don't worry, though. If you're a beginner, you should start with low-maintenance animals like chickens or rabbits. You can work up from there as you gain experience and confidence.

- **Desired products:** What goals do you have with livestock? Do you want eggs and meat? Are you interested in making milk, cheese, or even fibers? Some animals offer a few products, whereas others can be extremely versatile, so choose breeds that meet your needs and preferences.

Typical Off-Grid Livestock Options

In this section, we will cover all the animals that you would find on most farms or homesteads and a summary of them.

Poultry

- **Chickens:** One of the most popular options, they are popular for their eggs and meat. Choose the breed of chicken based on your desired level of production, temperament, and climate.

- **Ducks:** Ducks will also provide eggs and meat, and they will be a natural pest control method against slugs and various other insects. Find a breed that is suitable for your climate and that you have a nearby water source for them.

- **Geese:** Yes, you can have geese, and they provide great weed control and meat. However, you will need a pond or a larger body of water for them to access.

Small Livestock

- **Goats:** This is one of your most versatile animals. Goats will offer milk, meat, and mohair (fiber). They also provide land management with brush-clearing. Choose your breed based on their milk production, meat quality, and fiber characteristics.

- **Sheep:** Sheep are an excellent source of wool and meat, and some breeds will even produce milk. Choose breeds that are suitable for your climate and land while factoring in wool type and meat production.

Large Livestock

- **Cows:** These animals are great for producing milk and meat, and some breeds are great for grazing, which will give you land management. Cows will need significant space and resources, and you will need experience.

- **Horses:** These majestic animals need a high level of care and expertise, but they do offer transportation, hauling power, and companionship. If you have the experience and the land, you should highly consider adding horses to the homestead.

The important thing to remember is that each animal will have its own set of needs and challenges. This is going to require you to research the different breeds. You need to know their specific requirements, temperament, and even health concerns. Your choices should come down to the animals that are compatible with your climate, land, resources, and experience level. If you're still unsure about what animal to choose, consult with experienced homesteaders or breeders.

6.2 The Gritty Truth

Adding a feathered or furry companion to your homestead sounds like a great idea, but before you open your gates to an animal companion, it's crucial to understand the realities of off-grid livestock raising. This isn't the picturesque life that farm life is sometimes portrayed as; we have to look at the raw and honest truth. Some challenges wait for every homestead that incorporates livestock. This section is dedicated to those very real obstacles that will test any homesteader, especially those who aren't ready for it.

Physical Demands

There is no sugarcoated way to look at this. Raising livestock is hard work, and it's more taxing than anything else you will do on your homestead. From building and cleaning shelters to hauling feed around the property, you can expect regular, taxing physical exertion. Everything you do regarding your animals requires strength and stamina.

You must be prepared for long days, especially when you have animals with regular birthing seasons. There are also illness outbreaks and other unexpected emergencies that you will need to account for. Remember, you are responsible for the well-being of a living creature. This isn't work that you can "put off until later," so you should be ready for the physical toll that this will take.

Time Commitment

An underestimated aspect of raising livestock is the time investment. This isn't just a few hours out there feeding and watering. Animals need to be cared for daily, which means this is a task that doesn't care about the weather or your schedule. Early mornings are going to become part of your routine, as are late-night checks when the weather gets rough or it's a birthing season. You can still have time away if you so choose, but that means you will need to make arrangements for their care. That part is less likely on off-grid homesteads, but it is still something that needs to be considered.

Unpredictable Weather

Nature will throw some wild weather your way, which you will already know when you move onto the homestead. However, you now have to think about your animals, who are essentially on the front lines. Floods can threaten their shelter. Scorching summer heat requires shade and extra hydration, and the bitter chill of winter will require its own provisions. You should be ready to adapt your routine to that, as you may need to build an emergency shelter and provide comfort in any circumstances.

Predator Threats

You and your animals aren't the only ones in this remote area. You're also sharing that space with predators. Foxes, coyotes, raccoons, and even predatory birds will pose a threat to your animals. You'll have to think about your fencing and shelters, which offer more protection from predators. This also means that you need to be extra vigilant, whether with your own eyes or solar-powered cameras. There are also other preventative measures, like deterrents and guardian animals. But the one aspect that you will need to be most prepared for is the emotional toll that a loss can take when a predator is successful.

Financial Ties

You will seriously have to think about money here. The costs of starting to raise livestock can be steep. You're buying animals, materials for shelters and fences, feed, and other supplies. Then, you have the ongoing expenses that will stack on top of that with regular feed purchases, veterinary care, bedding, and necessary repairs. You then have to factor in the unexpected vet bills and potential animal losses. All of this will be a significant hit to your budget, and that's why it is crucial to plan carefully, manage your resources, and prepare fully for the financial commitment.

Planning and Preparation

These are living creatures, which means you can't rely on spontaneity. Instead of winging it when you bring your first animal onto the homestead, do thorough research and preparation. Learn about your chosen species. What are their health requirements? What are the common challenges? From there, you'll need to think about designing and building appropriate shelters for them. You also can't forget fencing and a comprehensive feed plan. The more you prepare, the easier this journey will be for you and your animals.

6.3 Animals Need Sanctuary

Livestock, whether you choose cows, sheep, goats, or horses, deserves a safe and comfortable homestead, just like you. These animals will need access to shelter, clean pastures, clean water, and waste management. When you understand and meet their needs, you not only ensure their well-being, but you also ensure that you are sticking to sustainable, ethical practices.

Building Their Haven: Shelter Essentials

Let's think back to the gritty truths. Your animals potentially face inclement weather, a hot sun, and potential predators. A shelter for them is a haven, giving them protection and a sense of security. So, what are some key aspects of a proper shelter?

- **Size:** The appropriate shelter size is going to depend on the number of animals and their species. There needs to be sufficient space for all of them to lie down comfortably without being crowded. Remember, each species will have varying needs. For example, sheep and goats will require a small shelter, whereas cows will need bigger shelters.

- **Ventilation:** Giving their shelter proper ventilation is vital for maintaining good air quality and preventing costly respiratory issues. This ventilation can come in the form of windows, vents, or leaving a side open to allow air circulation.

- **Elemental protection:** This shelter should offer protection from rain, snow, and wind. Material consideration is key. Wood, metal, or even concrete can withstand the elements. You'll need to ensure that the roof is strong enough for any weather extremes and provides enough shade in the summer.

- **Predator security:** Think about what predators are in your area. This will help your shelter design by incorporating sturdy walls and fencing to deter those predators from hurting your animals. You should also consider features like an escape route that your animals can use if there is an emergency.

Sustainable Pasture Management

A pasture is more than just an open area for your livestock to graze; it is an important ecosystem that is teeming with life. This means that you have to utilize sustainable management practices to ensure healthy pastures that will feed your animals while protecting the environment. The following are some key strategies you should use:

- **Fencing:** Pastures need strong and well-maintained fencing to set boundaries and keep your animals from wandering off your land. You will need to think about materials that are suitable for your animals and the terrain. Your other considerations are going to be durability, visibility, and how easy they would be to maintain.

- **Rotational grazing:** Instead of one large area, divide your pasture into sections and let the animals rotate through them regularly. By using rotation, you're even allowing grazing, preventing overgrazing, and giving these sections time to rest and regenerate.

- **Forage cropping:** You should plant different species in specific areas and allow your animals to sequentially graze on them. This is going to provide your animals with diverse nutrients while improving pasture quality. It also reduces the strain of various parasites.

Ensuring Accessibility to Water

Fresh, clean water is a must when you want to keep your animals healthy. You'll want to ensure constant access to clean water sources and that this lasts throughout the day, regardless of weather conditions. There are some very easy options for you that you can consider when designing your homestead.

- **Automatic waterers:** These can provide a continuous supply of water, and you'll minimize your worry about contamination.

- **Troughs:** These are great, but keep in mind that you will need to regularly clean and refill them, especially when it's hot.

- **Wells and pumps:** You can put a spigot near the pasture that will let you easily maintain your troughs or ensure that the automatic waterers stay functional. Ensure that you perform these proper maintenance tasks and test the water regularly.

Waste Management

Responsible waste management is so important not only for your animal's health but also for the protection of the environment. The following are essential practices that you will bring to your homestead.

- **Manure removal:** You'll need to regularly remove manure from shelters and pastures to prevent buildup, which will lead to contamination.

- **Composting:** With animals, you will have a wealth of manure that can be turned into valuable compost. This will be peak sustainability, and you'll be promoting soil health.

- **Wastewater management:** Properly handle the wastewater from these areas to prevent water pollution.

***

Everything I have covered with you is just pieces of the foundation. You will need to consider your location, climate, and animal breeds, and you need to give your animals the sanctuary they need. Don't forget that when your animals are happy and healthy, they will be highly productive.

6.4 Animal Care Essentials

When you think about your homestead's animals, you will slowly see that they are important members of your ecosystem. These animals are playing just as much of a role in your homestead's sustainability and well-being as you are. Therefore, we come back to the knowledge that their care is important. In this section, we will discuss the most essential components of taking care of your livestock.

Daily Routines

Care for your animals starts with the establishment of consistent routines. We've already visited these daily chores, but let's review.

- **Feeding:** You should research and provide each animal with the appropriate diet. Their diet will be based on their breed, age, and activity. Do your best to use homegrown feed, but supplement with store-bought feeds when you need them. Put your animals on a regular feeding schedule to make sure they have consistent access while avoiding things like overeating or competition.

- **Watering:** Your animals need constant access to clean, fresh water. Setting up a reliable watering system is crucial; therefore, you will need automatic waterers or a strategically placed trough system. If using troughs, ensure that they are cleaned and filled throughout the day.

- **Shelter:** While your shelters should be durable and well-ventilated, you need to do daily checks on them. Check for damages that could potentially let predators in. Then, you need to make sure that the shelters are cleaned to promote hygiene while preventing diseases.

Understanding the Unique Needs of Your Animals

Something that should always live in your head is that each animal species will have its own needs and considerations. If you have chickens, do you have an area for them to take dust baths? If you have goats, do you have structures for them to climb so they can get proper exercise?

This is where knowing the specific requirements will come into play. When you know their natural behaviors, social needs, and health concerns, you will be able to give them the care they deserve.

Preventative Measures

You don't want to wait for illness to strike your homestead. Conduct basic health checks on your animals, keeping watch for signs of illness or injury. Some basic things you can do are monitor their activity levels, appetite, manure, or overall demeanor. When you can detect problems early, you can intervene, preventing something minor from escalating into a major problem.

Vaccinations and parasite control are also measures that need to be implemented. Talk to your local veterinarian to determine an appropriate vaccination schedule and to establish parasite control methods that factor in the animal species and your location. The more proactive you are, the more you can safeguard your animal's health and well-being.

Building Trust and Human Treatment

Animal care, when you're off-grid, goes beyond basic care. You have to build trust and foster a positive relationship with all of your animals. This drives their emotional well-being, which drives their productivity. The following are just some ways you can build those relationships:

- **Gentle handling and interaction:** You will spend time with your animals doing basic chores, but you should spend time with just them. Give them treats or pets, or just gently talk to them. This will establish trust, which will make them feel more comfortable with you around. It's that comfort that will make handling and care much easier.

- **Respect their natural behavior:** Let your animals be in their most natural state as much as possible. Give them enough space to roam, forage, and socialize. This lowers the animal's stress levels and promotes their overall well-being.

- **Humane treatment at all times:** Although they are not humans, your animals are sentient, and they deserve respect and compassion. Their living conditions need to be comfortable, and they should be able to thrive just as much as you. Make sure that all of your livestock practices do not cause unnecessary pain or suffering.

***

Everything on your homestead is going to be interconnected, which means the health and well-being of your animals will directly impact the land and your homestead's livelihood. Healthy animals will produce nutritious food and mature for fertilization, land management, and even pest control. They keep the land healthy, and healthy land gives them the resources they need to thrive. It's a prime example of sustainability.

6.5 DIY Projects for 1000 Days of Animal Care

These projects will add functionality and charm to the homestead while saving money and reducing waste. We will tackle the hay feeder, natural fencing, and a chicken tractor. As usual, be safe when working with these materials.

Project 1: DIY Hay Feeder

This project is great for your smaller animals, like sheep and goats. It provides them with clean, dry hay while also cutting down on the chances of overeating.

Materials

- Wooden boards (various lengths and thicknesses)

- Galvanized nails or screws

- Hinges (optional)

- Mesh wire (which can be chicken wire or hardware cloth)

- Staple gun or zip ties

- Wood stain or sealant (optional, but a good idea)

Instructions

1. You will design and measure the feeder. Plan its size based on how many animals you have and their typical hay consumption. Sketch out the design with access points. Consider adding a roof with an overhang for weather protection.

2. After you have your design, cut your boards according to your design. You can assemble the frame using your nails or screws. Create sturdy walls and a slanted roof. You can use your hinges on the roof to make refilling the feeder a breeze.

3. Once your structure is sound, you can attach your mesh wire to the front and sides of the feeder, allowing access while cutting down on waste. You can secure the mesh with your staple gun or zip ties.

4. Apply a wood stain or sealant for weather resistance. This will keep the wood from splintering. Just make sure that all your materials are safe for your animals.

Project 2: *Natural Fencing*

While this is a project that takes time, given that you're using live materials, it's a sustainable way to define your boundaries. I would suggest using more budget-friendly fencing until your saplings can mature a bit.

Materials

- Live saplings (willow, hazel, and dogwood)

- Wooden posts (cedar or treated woodwork best)

- Natural twine (rope can be used too)

- Pruning shears

- Shovel

- Handsaw

Instructions

1. Plan the fence line and mark the desired boundary. You should consider your animal's needs and aesthetics. There also needs to be proper spacing between the posts for the saplings to grow.

2. You'll then need to prepare the posts. Dig holes for the posts at the appropriate intervals, making sure they will have sturdy placement. Set your posts and pack the soil back around them.

3. It's now time to plant the saplings. You'll want to choose saplings that are suitable for your region and climate. These saplings will go near the posts, but make sure you leave enough space for future growth.

4. As your saplings grow, you will be interweaving their branches horizontally between the posts. In the beginning, you'll use your twine (or rope) to secure the branches together, and this is what establishes the living fence.

5. As the saplings grow, you will need to regularly prune the excess branches, which will encourage dense growth and keep your fence at the desired height. Water the saplings regularly, especially as you are getting the fence established.

Project 3: Chicken Tractor

This moveable chicken coop is not only great for your feathered companions, but it will also help promote healthy soil through natural fertilization. This is a great project for more advanced homesteaders.

Materials

- Wooden boards and framing lumber (treated or cedar)

- Chicken wire fencing

- Hardware cloth (for predator protection)

- Hinges and latches

- Wheels (optional, but makes mobility much easier)

- Nesting boxes

- Feeder and waterer

Instructions

1. You will start by building the frame. It should be sturdy and use secure connections. Consider designs that implement a slanted roof to protect your chickens from the weather.

2. Cover the frame with chicken wire. You can use a staple gun for this as long as you ensure a secure attachment. The hardware cloth will be attached to the bottom of the frame to deter burrowing predators.

3. Create access points with hinged doors. This will give you easy access to clean the coop. At this stage, you should also consider attaching wheels. You can even find ones with locking mechanisms, but this will give you effortless movement between grazing areas.

4. You'll move on to the interior setup by installing nesting boxes, the feeder, and the waterer. This is a great time to ensure that your chickens have adequate space and comfort.

After this, you will just let your chickens roam. You can craft some expedient fencing to keep their grazing area contained and just move that and the tractor regularly to give them fresh grazing areas and promote soil fertilization. You get fresh eggs and healthy soil, while your chickens get room to roam and a cozy, safe coop.

***

Now that we have the basics of your off-grid homestead ready to go, it's time to think about your home. Moving off-grid isn't a turn back to the days of "roughing it" or living like pioneers. It can be a life that you feel truly comfortable in, and all you need is a little bit of natural power.

Chapter 7:

Harnessing Natural Power

Imagine a brisk morning on your homestead. The sun is rising, bathing your house in light, and the sound of birds is playing on beat with the sound of wind turbines. You look to your side, and your laptop is playing some morning music. That modern luxury isn't being powered by a distant power plant, though. It came to life thanks to the sun that beats down on your solar array. This isn't a far-fetched vision; it's becoming a common occurrence for people and even communities that choose an off-grid lifestyle.

Much like the reason to move off-grid, the movement toward energy independence is driven by the desire for self-sufficiency and environmental responsibility. There have been growing concerns about climate change, energy costs have been on the rise, and dependence on conventional grids has driven people away to look for other sources of power. And with the advancements in solar, wind, and hydro technologies, the possibilities for self-sufficiency are possible and accessible.

One of the biggest misconceptions about off-grid living is that life gets reverted to pioneer times, that it's all kerosene lamps and roughing it. However, many of today's off-grid homesteads are enjoying the same modern conveniences that you enjoy right now. (Harler, n.d.) They have all the same appliances, internet access, and even air conditioning. However, these modern comforts are all powered by the sun, wind, or flowing water. Some homesteads rely on one, and some will use a combination of methods, but all get to enjoy the benefits of power without relying on a traditional power grid.

This chapter will be your guide to the growing movement. From a solar-powered cabin in the woods to wind-powered homesteads on the plains, we will navigate each energy source's possibilities and practicalities.

A word of caution, though, before we dive into this exciting DIY world: there are local, state, and even federal regulations on alternative energy systems. Knowing what the regulations are in your area is absolutely crucial before going on this off-grid journey.

7.1 Solar Power: Harnessing the Power of the Sun

When thinking of complete energy independence and a more sustainable off-grid future, you will probably start by looking at harnessing the power of the sun. Solar energy has become the most popular off-grid energy solution. There is a good reason for that, too, as it gives you clean, renewable power that severs your reliance on grids and fossil fuels.

Catching Rays: The Basics

You've likely seen someone using solar power, but how does it work? Sunlight hits the photovoltaic (PV) panels, which are basically semiconductors. The panels turn the energy from those rays into direct current (DC) electricity. That's close, but our homes and appliances primarily rely on alternating current (AC) electricity, so how do we get DC to become AC? That's where your inverters come in. The inverter will turn the DC electricity that comes from the solar panels into AC electricity. (*How Does a Solar Panel Work: Step by Step - Qcells North America*, 2024)

So, just find a solar panel, and you're good to go, right? Not really. Choosing the right PV panels is crucial. These panels come in various sizes and efficiencies. Higher-efficiency panels generate more power per unit area; therefore, you will need to factor in things like your climate, average sun exposure, and your energy needs.

Components and Configurations: A Deeper Dive

First, we will start by looking at all the pieces of your solar power system.

- **PV panels:** This is the main part of your system and what will capture the energy from the sun.

- **Inverter:** The next crucial component of the system is the one that turns the DC electricity taken in by the panel into usable AC power.

- **Batteries:** This is where the excess solar energy is stored, and it will be used at night or on cloudy days.

- **Charge controller:** This will regulate the charging process and is crucial for preventing damage to your batteries.

- **Mounting hardware:** Whether you're mounting your panels to your roof or the ground, you want to ensure they are secure.

From the basics, we need to think about system sizes and configurations.

- **Small system:** This method is great for powering lights and appliances. You'll often use this when you're still on the grid or when using backup generators.

- **Medium systems:** These will cover more of your needs. I would recommend this if you have an off-grid home but not too much land.

- **Large systems:** This setup will allow you to move off-grid without missing a thing from the world you know now. You have complete independence from municipal energy.

Just because this is a great off-grid resource doesn't mean you can't use these systems now. More areas are becoming open to small systems. These would be grid-tied systems. They connect to the utility grid, and they will allow you to sell the excess power, only drawing energy when you need it. These are becoming stellar ways to save money while still on the grid.

Using a grid-tied system will allow you to get hands-on experience before moving off-grid. Remember, once you sever ties with the grid, you are on your own. You will need to have sufficient battery storage for nights and cloudy periods.

Planning and Optimization: The Overlooked

Before diving headfirst into solar power, you need to conduct an energy audit. Knowing your daily and seasonal energy consumption is essential when it comes to choosing the right system for your homestead. Pay attention to your energy bills or ask your provider for a more detailed look.

When you choose your system, optimization is the next step in trying to achieve maximum efficiency. The following are ways that you can reach that efficiency:

- **Panel orientation:** South-facing panels are going to capture the most light. Watch shading patterns on your homestead and think about adjusting the panel tilt to keep maximum exposure all year long.

- **Batteries:** Battery choice is also important, and you will want to choose the most efficient technologies that you can fit into your budget. Lithium-ion batteries will allow for a longer lifespan and deeper discharge capabilities.

Remember, being off-grid means being energy-conscious. When buying appliances, choose the most energy-efficient options. Change your usage habits and learn the periods when you have a higher energy demand. These considerations will allow you to have sufficient backups when you need them.

The Pros and Cons

While solar energy is one of the most popular choices for off-grid power, there are advantages and disadvantages that you need to weigh out.

Pros

- **Clean, renewable energy:** This reduces or eliminates the reliance on fossil fuels, which is beneficial for the environment.

- **Reduced energy costs:** This is really a pro for grid-tied systems, but you will see a reduction in energy costs over time.

- **Energy independence:** When you move off-grid, you will find greater control and resilience.

- **Government incentives:** With the newfound openness to solar power, many regions also have financial incentives for installing solar panels, whether off the grid or on.

Cons

- **High upfront cost:** Despite the long-term benefits and the potential incentives, the initial investment in solar energy is steep.

- **Weather dependence:** You have to have sunlight, and some regions can have periods of cloudy weather and extremely short days, which calls on the need for another source of energy.

- **Battery storage cost:** After your initial investment, you're still going to need battery storage. Batteries are going to be expensive if you choose high-end technologies, and you can't forget large systems where you're completely severed from the grid.

Solar energy is going to give you the greatest path toward independence. When you can carefully consider your needs, optimize your system, and maintain energy-conscious habits, you will be able to do so with ease.

7.2 Wind Energy: Harnessing the Power of a Breeze

Much like the sun in the sky, the wind is also a powerful and renewable resource that you can harness. Wind energy, the process of taking the kinetic energy of the wind and converting it into electricity, gives you a clean and sustainable alternative to traditional fuels. But before you start raising turbines on your property, we should take a minute to dive into this world a bit and figure out what you need to tap into this natural phenomenon.

Catching a Breeze: The Basics

You're likely familiar with these machines. They are like giant pinwheels catching the wind, but instead of a children's toy, these are majestic giants with their sleek blades slicing across the sky. As the wind spins the blades, the rotation moves down the shaft, turning the kinetic energy of the wind into usable electricity. The conversion takes place with the generator, which is the heart of the turbine.

Keep in mind that not all turbines are equal. Wind turbines will come in various designs and sizes, but it really comes down to type. You have horizontal-axis wind turbines (HAWTs), which are what you've seen more of on wind farms. There are also vertical-axis wind turbines (VAWTs), which have their blades rotating around a vertical pole. Knowing which one to choose is going to require you to know your wind pattern and terrain well before moving off-grid (Clayton, 2021).

- HAWTs are great for open spaces with consistent winds, and their long blades are great for capturing higher wind speeds. (Clayton, 2021)

- VAWTs are better in turbulent and low-wind areas because of their omnidirectional design (Clayton, 2021).

If you are thinking more small-scale or using wind as a supplemental power source, micro-turbines are also a great solution. These compact designs fit on rooftops or small poles, and they are certainly a DIY project that you can use around the homestead.

Diving Deeper

Now, we can dive a bit further into the intricate workings of your wind system. Let's start with your key components.

- **Tower:** This structure supports the turbine. It has to be sturdy, and it has to be high enough to reach the wind.

- **Inverter:** Like with your solar system, the energy generated is DC electricity. The inverter makes that into AC electricity, which is compatible with your home.

- **Batteries:** These will store the excess electricity generated during windy periods. You then get to tap into this when the wind is low.

- **Home wiring connection:** This will be used in any of your energy solutions, but since there is DIY potential with turbines, this component is what delivers the electricity to your home.

But you can't just get a turbine and have power. You have to take a crucial step long before you move off-grid, and that is conducting a wind resource assessment. This is essentially a detailed weather report solely focused on wind. By measuring wind speeds and directions over a long enough period, you can determine if your location is going to have enough wind power to make a turbine feasible. It will also help you choose the right size and type of turbine for your homestead.

Also, keep in mind that harnessing wind power comes with its own set of considerations. Permit regulations will vary by location, so you will have to adhere to those rules before setting up your off-grid wind farm.

You will need to consider the height of your tower and understand that you will need to ensure its stability. There are also safety measures like lightning protection that need to be in place.

The Overlooked Aspects

The wind is a more fickle ally than the sun. The intensity fluctuates so often, which means your power output is going to fluctuate, too. You should be ready for the times when the wind is more of a whisper than a roar, which means looking into backup sources like a generator or a small solar system. This will ensure you get through the less windy times with uninterrupted electricity.

You also have to factor in noise. Modern turbines are significantly quieter than they have ever been, but larger models still have the potential to produce a noticeable low hum when nearby. You'll need to think about your noise sensitivity and look into noise regulations before your turbine goes up.

The last aspect will be in the aesthetic department. A towering turbine will definitely clash with a lot of landscapes. You will have to take time to plan and consider your surroundings to ensure that your turbine complements the environment.

The Pros and Cons

Wind energy will need to be considered a bit more in this category, as it brings its own share of advantages and disadvantages.

Pros

- **Clean and renewable:** Wind energy has no harmful emissions, which is key to combating climate change. For regions with an abundance of wind, it's a solution that will always be there.

- **A great on-grid solution:** Much like solar power, you might be able to install a turbine at your current home. This will allow you to get much-needed practice with alternative energy while saving on your energy bills.

- **Economic benefits:** Maybe you have a lot of open property. This could be the sight of a potential wind farm, which will generate income with land leasing. This one is a bit more of a stretch, but the potential is there.

- **Technological advancements:** Turbines are becoming much more efficient and affordable. Again, in the right region, wind power is becoming an increasingly attractive option for homesteaders.

Cons

- **Intermittent resource:** Even if your region has an abundance of wind, those wind speeds are going to vary, which will lead to fluctuations in your turbine's power output. You will need a backup source if you want uninterrupted power.

- **Initial investment:** Wind turbines can be expensive. However, these costs are decreasing, and incentives are available for homesteaders.

- **Visual impact:** Larger turbines will begin altering the landscape, which brings up aesthetic concerns. This is something to consider if you are living closer to a community.

- **Potential noise issue:** Again, modern models are running quieter than before, but you still might get a noticeable hum. Noise regulations and distance considerations will also need to be made.

While solar will be the dominant choice for off-grid living, wind energy provides a great solution. You might even consider using a small turbine with a medium solar system. This is why knowing your land, the regulations, and your weather conditions is so important before moving off-grid.

7.3 Hydro-Power

When you move off-grid, you need to know what all your resource options are to truly harness the power of nature. While solar and wind are the most common systems found in off-grid situations, hydropower has emerged as a captivating option for some homesteaders. As we have done with the other solution, let's look at the basics, intricacies, and considerations of this renewable resource.

The Basics

This works like wind energy, except with water. Flowing water sources will power a turbine. That turbine will drive a generator, which turns the kinetic energy into electricity for your home. For the off-grid homestead, there are two main systems that you can choose from.

- **Run-of-the-river:** These systems use the natural flow of a stream or a river without needing a dam or other watercourse alteration. These systems are great for moderate water flow and heat (the vertical drop of the water), but they tend to generate smaller amounts of electricity.

- **Micro-hydro:** These systems will operate on a smaller scale as they often use gravity-fed water sources like springs or creeks. These are great for homesteads with limited water sources. However, these will require careful planning and design to get the most efficiency.

Before putting a system in a water source, though, you will need to face things like regulations and environmental considerations. You will face things like water rights and permitting processes, but you also have to consider the potential environmental impact that you will have on the ecosystem.

Components of the System

While the concept of hydropower is straightforward, there are a lot of pieces at play, so let's look at those:

- **Turbine:** This is the heart of your system, and it works just like a wind turbine. Different types of water turbines also exist, and each of them is suited to specific water flow and head characteristics. This is why careful planning is crucial.

- **Penstock:** This is the pipe that channels water from the source to the turbine. It's an efficient way to ensure water delivery to the turbine.

- **Generator:** Driven by the turbine, this is what turns the mechanical energy from the turbine into electricity.

- **Controller:** This regulates the system to make sure operation remains safe while also optimizing power output.

- **Batteries:** Like with your other energy solutions, excess energy is stored here to be used during low-flow periods.

The Importance of Water Flow and Planning

Without water flow, you have no hydro system. Much like you would conduct a wind assessment, you will need to conduct a thorough water flow assessment. This measures the flow rate and head throughout the year, which will tell you the potential energy generation of the source. From there, you can determine if hydropower would be feasible.

After the assessment, choosing the right turbine size and type is also critical. Oversizing can lead to inefficient operation, and under sizing can stifle your power output. You will have to consult with a professional to make sure your system matches your resources and needs.

Beyond the Basics

Like any renewable resource, some considerations need to be made.

- **Variability of water flow:** No water flow will remain consistent forever, which means you have to be ready for periods of low water flow. Having a solar or wind setup will be a great backup.

- **Maintenance:** You want to ensure long-term system operation, so regular maintenance is a must. This means removing debris, cleaning out sediment buildup, and ensuring you have safe access to your components.

- **Environmental impact:** Respecting the existing ecosystem is a must. Your hydropower system must have little impact on water flow, fish habitats, and the natural beauty of the area.

The Pros and Cons

Pros

- **Clean and renewable:** Hydropower stays in line with your sustainable lifestyle as you generate electricity without harmful emissions.

- **Reliable:** With proper maintenance, hydro systems can provide a consistent source of power.

Cons

- **Site-specific:** You have to be near a water source, and it has to have consistent water flow and head.

- **Regulations and permitting:** The legal requirements for hydro systems can be complex, time-consuming, and expensive.

- **Environmental impact:** Hydro systems require a little more planning because they still have an impact on the ecosystem; therefore, you have to do your best to minimize this.

Hydropower can be a great tool for off-grid homesteads. If you manage to land water rights and have a good source of water, then there is no reason not to use this resource. However, it's something that takes a lot of research, planning, and respect for the environment. I would also recommend using it alongside another renewable source to ensure there are no interruptions to your electricity (Meissner, n.d.).

7.4 DIY Project for Small Energy Generation

When it comes to powering the homestead, you'll have to think much bigger. While you can technically set up these systems yourself, many of the components are already made and ready to go. With that being said, there are a few projects that you can do to generate a small amount of electricity, which can help charge your battery storage. I will say, though, that knowing at least one of these projects will give you a better idea of the inner workings of your much larger systems. For that insight, we will turn to a water turbine generator.

Capturing the power of moving water with a turbine generator you made yourself is a thrilling experience. While its uses will be limited, it can perhaps power a motion detector that shines on your property, or you can connect it to your battery bank.

Materials

- Bicycle: This is the backbone of this turbine project. You should look for a model that has a sturdy frame and removable wheels.

- Alternator: You should be able to find one of these in any car junkyard. This is what will convert the mechanical energy into electricity. Keep in mind that you want one that is in good condition.

- Plastic Cups or Balls: These are going to be your water scoops, which are the part of the system that actually spins the turbine. Choose materials that can hold up to moderate water currents.

- Screws, drill, and screwdriver

- Saw and welder: You don't need to be an expert welder, nor do you really need one. However, it will make attaching the alternator simpler.

- Clamps: This is a great alternative for those who don't have a welder.

Instructions

1. Simply enough, you'll remove the front wheel and loosen the chain.

2. Now, the tricky part is going to come when you mount the alternator. Again, welding will simplify the process, but if not, you'll need to find a way to securely mount it to the underside of the pedals. You want to ensure that the chain will wrap around the alternator's pulley. If you're feeling inventive, you can replace the pulley with a bicycle sprocket. The sprocket will give you a much better fit and increase your success rate.

3. Adjust the seat to where it is at its full height. This will maximize the distance between your alternator and the water, preventing it from being submerged.

4. Now it's time to make the turbine "blades." Cut several plastic cups or balls in half, creating your mini-turbines. Attach them to the bike's rear wheel with screws. Try to keep them spaced around 2 inches apart and facing the same direction to be effective. The best rule of

thumb is to make sure that the flowing water will turn them clockwise, which will match the alternator's rotation.

5. The last thing you need to do is get the bike into the water. Make sure you're setting it upside down in a flowing stream or creek, with the alternator well above the water line. The current should push against the blades, causing the wheel to turn, which should bring the alternator to life.

If everything goes the right way, you are generating electricity, which means you now have to put it somewhere. Use the appropriate wiring to connect the alternator to the batteries on the shore. Keep in mind that the strength of the current will depend on water flow.

Tips and Tweaks

For a stronger current from the alternator, use hard plastic ball halves instead of plastic cups. This will give the system more resilience. You can also consider using multiple small water turbines, which will give you more power. Also, don't be afraid to explore using various materials and blade designs to optimize the system. Again, it's not going to power your home, but there might be other applications for those batteries.

When working with a DIY system like this, you need to prioritize safety. You're working with electricity and water, so ensure proper insulation and avoid submerging the alternator. You can even add plastic guards around the alternator to protect it from water, but the most important thing is not to take unnecessary chances.

***

The homestead is now powered! We will come back to this toward the end of the book, where you'll learn how to combine your power sources and establish a localized grid. But let's switch gears to an important aspect of off-grid living. When you're living away from people, you might feel like there are eyes on your homestead, waiting until you're not looking to take advantage of your hard work. That's when homestead security becomes our next priority.

Chapter 8:

Off-Grid Home Security

Picture this: You're standing on the porch of your off-grid home, watching the sun sink behind the tree line. There is a moment of tranquility as you think about how far you have come in achieving your dream of self-sufficiency. But as the last bits of sun disappear in the sky, something stirs inside you—an uneasy feeling. The isolation that you had just been appreciating now bears a different weight. The comforting silence becomes too much as it amplifies the rustle of leaves in the distance. Suddenly, the night seems to be filled with eyes, all fixated on your homestead.

Living off-grid brings about so much freedom and connection with nature, but it also presents some unique challenges when you think about security. There are no flashing blue lights to deter unwanted visitors or blaring alarms to call for help. You are your first line of defense, and while your homestead holds endless beauty, you can't rule out potential threats.

Unlike the neighborhoods you're used to, relying on your neighbors to keep watch isn't an option. Even in a more rural area, your neighbors are sometimes a mile away, if not more. Isolation becomes normal, and you are bound to have more than one encounter with danger. Bears could rummage through your trash cans or waste piles, raccoons could be eyeing your chickens, and the occasional fox could be hanging outside your pasture, stalking your livestock. These natural dangers are a concern, but while you may have already accepted that those are a part of nature, you have likely forgotten that you have solar panels, exposed generators, water storage tanks, and more. All these things become tempting targets for someone looking to score on your hard work.

This is very much your call to action and a reminder that securing your off-grid haven needs a far different approach. Fortunately, you don't need to plan your moats and drawbridges. It's all about layering strategies, from physical barriers to off-grid technological solutions. With layering, you'll create a security system that's just as resilient as your homestead.

This chapter will explore some of the vulnerabilities that you'll experience when you go off-grid, from deterring wildlife to securing your resources. By the end of this chapter, you won't just feel safe in your home; you will feel empowered. You'll have the knowledge and the tools that will give you the safety and security you need, and you'll be able to watch the sunset without those feelings of uneasiness.

8.1 Building Layers of Protection

It can be easy to overlook security when you go off-grid, and when you do think about it, it can seem daunting. A lot of that challenge comes when you are no longer relying on the infrastructure and immediate response of your neighbors and police. But all you really need is a new approach. When it comes to your off-grid homestead, you need a proactive approach that will deter and delay potential threats, which gives you ample time to react. This is where the "bullseye method" will come in, guiding you to multiple layers of protection. These layers will come together to give you peace of mind, knowing that your homestead is secure.

Layering Your Defenses: Starting at the Perimeter

With traditional neighborhoods, there are several layers to the bullseye method. However, the one thing that remains the same is that your home is at the very center of that bullseye. When you move off-grid, your first layer lies at the very edge—the layer that encompasses your entire property. This is where fences are going to play a major role. They will act as both a physical barrier and a psychological deterrent. You'll have a few options when off-grid, so let's look at some of the friendliest to you.

- **Metal mesh fences:** These fences are strong and durable. They deter climbing, and they give you clear visibility for monitoring the perimeter. Your best bet is galvanized steel, which will have increased weather resistance. You might also consider using barbed wire as an additional deterrent, especially in high-risk areas, like ones that aren't in plain sight from your home.

- **Woven wire fences:** This is a budget-friendly choice for larger perimeters. They are also a great choice when reinforced with tension bars, as this adds strength to the material. Opt for tight weaves to prevent small animals from getting out.

- **Living fences:** Plant densely growing hedges like hawthorn or holly, which will give you a natural barrier and some aesthetic appeal. These fences are great, but they need regular maintenance and might not offer the same level of security.

Remember, the success of your fence goes beyond your choice of materials. Proper installation is crucial with fencing. Your posts should be spaced no more than 8 feet apart, and you should use secure ground anchors (typically concrete or metal posts sunk deep in the ground). Aside from the posts, you will need to make sure there is proper tension in the material, which will create a sturdy barrier.

You'll also need secure access points, which means you need to install sturdy gates that are equipped with high-security padlocks that are resistant to picking or other tampering. You can also install solar-powered automatic gate openers, which will allow some added convenience and control, especially if remote access is required. Of course, don't forget to add deterrents like "No Trespassing" signs. Visible security cameras (even if they aren't live) will add another layer of discouragement to unwanted visitors.

Lighting the Homestead: A Bright Deterrent

Intruders like working in the cover of darkness, which means nighttime on your homestead can be their ally. You can combat this by placing motion-sensor lights around your perimeter. You can incorporate a mix of floodlights and spotlights to light your pathways, potential hiding spots, and entry points. Your best friend here will be opting for solar-powered models for compatibility with your off-grid lifestyle and long-lasting operation. You might also consider dusk-to-dawn lights near your house for additional security and ambiance.

Take your surveillance to the next level with trail cameras. These discrete devices record videos or images when motion triggers them. This lets you monitor those remote areas that you can't always see. Not only can they identify suspicious behavior, but they can also help you track any wildlife that visits your property. You can place these cameras strategically along your property lines, near game trails, and even around other structures around the homestead. Another suggestion would be to invest in cellular-enabled models. These can give you real-time alerts and remote monitoring capabilities.

Securing the Core: Eyes on Your Home

With your homestead, you can break it down into even finer layers, which can include your home's fenceline, the entry points, and all the way up to your safe room (if you have one). You can also keep it broad, but the important thing to keep in mind is that off-grid living can potentially mean much slower response times from law enforcement. That is why home security when you're off-grid becomes critical.

Doors and windows become prime targets, so to combat this, invest in solid-core doors and install metal frames if possible. Your locks should also be upgraded to deadbolts with at least a one-inch throw. For those, I would suggest looking for the types with high security ratings. You could also use smart locks for keyless entry and remote monitoring. To further your door security, you should use reinforced hinges and strike plates to ramp up your defenses.

Now, let's look at your windows. I would recommend using security bars or shutters. Bars give you permanent protection, and they can be installed on windows and doors. The only concern is with the aesthetics of the bars. Shutters, on the other hand, can give you more flexibility. You can close them when you're away or any other time you need, and they can add a unique aesthetic to the home. Again, choose the options that are rated for security and durability, which are typically your steel or reinforced aluminum options.

You can't forget your outbuildings, either. These are just as crucial as your home, so install tamper-proof locks on sheds, workshops, and garages. You can add solar-powered alarm systems to these structures, especially if they are housing valuable supplies and equipment. Again, systems with remote access and notification capabilities will alert you to any suspicious activity, even when you're away from the homestead.

Using Landscaping for Security: A Boost from Nature

Off-grid is all about being resourceful, which means your security doesn't always need to be a manufactured solution. Nature can be a powerful ally. Look at your property lines and vulnerable areas (areas not in your line of sight from your home); you can plant thorny shrubs like roses and barberries in these places. Dense foliage like evergreen trees or claiming vines can create obstacles, which will hinder any access from unwanted visitors. Even around your doors and windows, you can plant things like holly or pyracantha to create a thorny or prickly obstacle. If you plant anything near your house, though, you will need to maintain your landscape to ensure you have clear sightlines to as much of your property as you can. In short, don't create hidden access points that predators can exploit. Keep areas pruned and ensure that your walkways are clear of any obstructions.

Fortifying Your Home's Defenses

Beyond the realm of doors and windows, these are just some other things that you can consider to strengthen your home's defenses.

- **Security screens:** You can install reinforced security screens on your more vulnerable windows for extra protection without giving up any ventilation.

- **Window film:** There is also a shatterproof film that you can put on windows. This is ideal for your larger windows, and it will deter any smash-and-grab attempts.

- **Door reinforcements:** You can use door jamb reinforcement kits to make your door frames even stronger, which prevents forced entry.

- **Alarm systems:** You can always install a solar-powered security system with motion sensors, door and window contact sensors, cameras, and siren or remote notification capabilities. You can do something as simple as a ring system, or you can spend a bit extra to get a system with professional monitoring for more peace of mind.

- **Storing your valuables:** You should put your valuables and important documents in a fire-resistant and waterproof safe. Anchoring this securely to the floor or the wall will prevent it from being stolen, too. The only drawback is the initial investment cost, but it's worth protecting the things that mean the most to you.

Building Layers of Awareness

Physical barriers are crucial, but that shouldn't be where your defense preparations should stop. Never underestimate the importance of awareness and preparedness. The following are more ways you can ensure that you and your home stay safe.

- **Be observant:** You want to know your surroundings as well as yourself. When you do obtain that understanding, you will notice if anything is amiss in your environment.

- **Develop a plan:** Much like you would as a disaster preparer, you should create a comprehensive emergency plan. This plan should outline what to do in case of a break-in, fire, or any other threat. You should also regularly practice this with everyone in the home.

- **Pets:** Your animals don't have to just be livestock; there is always responsible pet ownership. Dogs can be a natural deterrent, and they can alert you to suspicious activity.

- **Establish community:** This won't be for all homesteaders, but if you are living close to other people, you should connect with them. This will establish a supportive off-grid (or rural) community, and you can look out for each other.

- **Defense training:** These are just good skills to know, but they can be particularly helpful when you move off-grid. If you're comfortable with this, consider enrolling in self-defense classes or taking firearms training for more confidence and preparedness when off-grid.

Remember, home security is an ongoing process, not a one-time fix. Make time to assess your defenses regularly. You should always be aware of potential threats and ready to adapt your methods as needed. When you implement thorough planning, layered security measures, and an observant eye, you will establish a safe homestead for you and your family.

8.2 Off-Grid Technological Solutions

When you are in charge of your own security, you start to see the bigger picture—the one just outside of the connection to nature and self-reliance. This is what has led some homesteaders to establish a grid-like system for their comfort and security. In this section, we will take a closer look at the technological solutions that provide security features that closely resemble what you would be able to have while on the grid. This knowledge will then empower you to build a strong defense.

Keeping Eyes on the Homestead With Cameras

Solar-powered cameras feed off the power of the sun for continuous vigilance, and they will give you 24/7 recording without tapping into your homestead's power. This allows you to always monitor your property. When you are looking, you should choose models that have a long-lasting battery and weatherproof casings. There are a few other features that you might consider with your cameras, too.

- **Cellular connection:** Off-grid living shouldn't stop you from monitoring your property. Cellular-connected cameras will transmit footage and alerts, and you don't need wired internet or Wi-Fi. This will keep you informed about what's happening, even when you're away from the homestead. These cameras may cost a bit more, but they will give you peace of mind and the ability to act immediately.

- **Night vision and motion detection:** Off-grid living often means darker areas, which makes night vision and motion detection cameras so important. These cameras will capture activity in low light, ensuring that you don't miss anything unusual. Features like wide-viewing angles and infrared illumination will allow comprehensive coverage, and they are ideal when used on the perimeter of your property.

- **Live feed and alerts:** Being proactive is key when you're off-grid, and as you get closer to home, you will want to consider cameras that offer live-feed capabilities and push notifications. These alerts will go straight to your phone, allowing you to react in real-time. When combined with cellular connections, you will have even more reactive abilities. This is especially true if you choose a model with two-way audio communication. You can deter intruders and interact with visitors from anywhere.

Sounding the Alarms

- **Cellular connection:** Yes, we come back to this aspect of home security, but when you think about it, how effective would a blaring alarm or a bright light be when you're miles away from people? There is always a possibility of intruders not being phased by these things, especially if they have been watching you. Opt for cellular-based alarms, which will send a signal directly to you no matter if you're in your home, on the property, or away from the homestead. This lets you take appropriate action.

- **Motion sensors and tripwires:** Use tripwires and motion sensors to send up alarms. These can be more effective when used in restrictive areas, which will let you be aware of your entire perimeter. Always choose sensors with adjustable sensitivity and weatherproof casings.

- **Camera integration:** Visual confirmation will help tremendously when you're off-grid; therefore, you should integrate your alarm system with your camera array to get video verification whenever an alarm is triggered. This lets you assess what's going on, note false alarms, and give evidence to law enforcement if needed.

Creating a Smart Home Hub

- **Creating a centralized command center:** If you have established your own "grid away from the grid," you should consider a smart home hub. You might already be familiar with this, as it centralizes control over your lights, thermostats, and other devices. This hub can even integrate with your security systems. By managing everything remotely, you can add a new layer of deterrence. It will even allow you to lock your doors if you have smart locks installed.

- **Simulated occupancy:** You can deter unwanted attention on your homestead by creating the illusion that someone is home when you're not. Many appliances will have smart features; therefore, you can turn lights on or off, play music or TV sounds, or do anything else that makes it seem like someone is home.

- **Remote gate and door control:** This is ideal for extended periods when you are off the property. You will be able to grant access to trusted visitors remotely by using smart door and gate control features. Gone are the days when you had to leave keys hidden, which can be a risky move if the wrong person knows those hiding spots. When you purchase these systems, you should choose ones with secure encryption and other access control protocols.

***

Securing your off-grid home is something that you can do DIY for, which we will get to shortly, but for peace of mind, you might want to invest in more secure systems that will allow you to monitor the homestead. My last suggestion here is to get insurance for the homestead. You can never be too sure, and there are policies tailored for non-traditional living. Talk to an agent and get a plan that best suits your off-grid property.

8.3 DIY Projects for 1,000 Days of Home Security

Quality home security can come with a steep price, but if you're looking for a way to step up your security without breaking the bank, then these three projects are for you. Each of them will offer a layer of security that you can find peace of mind in. Keep in mind that while these are effective deterrents, they are not the complete security solution. Just think of them as helpful layers that can help your budget.

Project 1: Solar-Powered Motion Sensor Alarm

This is an eco-friendly alarm system that uses the power of the sun to deter any unwanted visitors to the homestead. It's fairly easy to set up, and it will provide a visible and audible alert when any motion is detected. I would suggest this setup for the perimeter of the property.

Materials

- solar panel with a mounting bracket (5–10 watt panel will do)

- motion sensor with alarm

- loud siren (or other noisy system)

- mounting hardware (screws, or zip ties)

- weatherproof enclosure (optional, but recommended)

Instructions

1. Start by choosing a location. This should be a strategic spot that gets plenty of sun. You can put it near your porch or entry points, or you can use it for those areas that might be out of your sightline. Ensure that the motion sensor's coverage area covers the desired zone.

2. Mount your solar panel first by attaching it to a stable surface. Use the provided mounting bracket to ensure its security. Also, make sure that it receives direct sunlight for the majority of the day.

3. Install your motion sensor. It needs to be within range of the solar panel, and the detection zone needs to cover the chosen area. You'll have to follow the manufacturer's instructions for mounting and angle adjustment, as there are various models on the market.

4. Connect the solar panel's output to the motion sensor's power input. Make sure you follow the provided instructions to ensure the system works.

5. Connect the alarm to the motion sensor's output, again, using the manufacturer's instructions. If mounting this close to the home, you might consider placing a secondary alarm inside your home for added protection.

6. Give the alarm a thorough test. You can do this by triggering the motion sensor, which will help you make sure that the alarm and solar panel are functioning. Once you're sure everything is good to go, then use weatherproof enclosures to protect your components from harsh weather.

Project 2: Tripwire Alarm System

If you're looking for a low-tech solution, this classic DIY project will give you a notification and the unwanted visitor a noisy surprise. Keep in mind that even a visible tripwire can have the power to deter intruders before they enter the area.

Materials

- fishing line (preferred, but you can use thin rope)

- bells or empty cans

- duct tape or zip ties

Instructions

1. Plan the layout. You want to implement some strategy and focus on potential points of entry around the home. Think about where an intruder would want to enter from, but avoid areas where you could cause an accidental trigger.

2. String your tripwire (fishing line) tautly across your chosen path. Make sure you have a way to identify these areas, especially when there is little or no light. Tie bells or empty cans to the wire at select intervals (preferably hidden). This is what will create noise when it is triggered.

3. Attach the tripwire ends firmly to a fixed object using duct tape or zip ties. You'll have to be careful here to ensure that the line is tight enough to make noise but not so taut that it breaks easily.

4. It will take some testing to make sure everything is just right, so walk through secured areas to test the tripwires. This is to test the tension and placement. Adjust where you see fit.

Project 3: DIY Security Camera With Recycled Materials

This project will test your innovativeness, and it repurposes readily available items to make a basic motion-activated camera. This will allow you to capture potential intruders on video. Now, this is DIY, so don't expect high-definition viewing; however, it will give you valuable evidence.

Materials

- old tablet or smartphone with a motion detection app

- sturdy container (shoebox will work; plastic container is a bit more durable)

- lens (magnifying glass lens or binocular lens)

- mounting hardware (screws and brackets)

- solar power bank (optional but allows extended recording)

Instructions

1. Start with preparing the container. Cut a small hole in the front of the container that accommodates the phone's camera lens. The lens of the phone's camera should align with that hole.

2. While ensuring the camera lens is positioned through the hole, secure the phone inside the container with tape.

3. Attach your other lens in front of the camera hole. This is going to give you better focus and clarity. You might consider trying different lenses until you find one that gives you optimal results.

4. If you haven't already done so, download the motion detection app. You may have to try a few out to find the one that works for you. Some will just take photos when motion is detected, but you want to find one with video capabilities. After you download this, you'll configure the app's sensitivity and recording duration to match your needs.

5. Connect the phone to a power source. In some instances, you might have a direct line, but in most cases, you'll need a power bank. This is where a solar bank will be your best option. Make sure that you have enough room to plug the charging cable in.

6. The last thing you'll need to do is test and secure the system. You can test it by triggering the motion detection system and making sure it records the video successfully. You'll use this time to adjust the app's settings and angle to make sure you get the coverage you want. Once you have the settings you want, close the container. Make sure that the phone has adequate ventilation.

Additional Tips

- You should customize all of these projects to fit your and your homestead's needs. This might require some research and comparing different options before buying components.

- Again, these options are DIY solutions and are not meant to be your sole source of security. Consider integrating them with your other security systems for optimal protection.

- Regularly test and maintain all your security systems, including your DIY solutions, to ensure they remain effective.

- Also, keep in mind that these projects act as deterrents, but there is no guarantee that the intruder will be stopped. This is why modern, sophisticated equipment is needed.

With these DIY projects, professional security advice, and an overall proactive approach, you can enhance the security of your homestead for greater peace of mind.

Now, we can take something that was hinted at here, along with our lessons in energy from earlier, and bring those forces together to establish what will set your off-grid homestead leaps and bounds ahead. In the next chapter, we will establish a local grid and power your way into a more sustainable future beyond the first 1,000 days.

Chapter 9:

Establishing the Localized Grid

You've secured the off-grid mindset, ventured off the grid, and carved your own path toward self-reliance. You've captured the power of the sun with solar panels, caught a breeze with your wind turbines, and maybe even set up a hydropower station to take advantage of flowing water. Of course, that means you have invested in quality energy storage to ensure that you have uninterrupted power every day, all year. But for this last chapter, we will take that a bit further to solidify your 1,000-day journey (and beyond) by establishing a localized grid.

Look back at some of the core values that drew you to the off-grid lifestyle. You have resilience in the face of potential disruptions, independence with your resources and power supply, and lessening the impact we've had on the environment. These things will be amplified tenfold with a local grid. Imagine, if you will, a widespread prolonged power outage. At a time when others could be scrambling for solutions, your homestead is thriving. You don't miss a second of comfort thanks to the shared power flowing through your grid. This interconnectedness further solidifies your resilience, giving you a greater sense of security and self-sufficiency.

Now think again about the energy solutions you might have implemented on the homestead. You might have a solar array, turbines, or a micro-hydro system. Many homesteaders will try to mix things up with two or all three solutions, which is a great start. But while these are powerful solutions, they remain on their own islands.

This is where the magic of a local grid comes into play. Think about uniting your independent energy generators and connecting them with a well-designed network. This will allow efficient power distribution, giving all parts of your homestead the energy they need. It enables optimized energy utilization, directing excess power from one source to fill gaps in another. With a grid, this surplus can be stored or shared, and as your needs change, your grid's inherent scalability will allow you to bring in new energy sources or even expand your home's power consumption.

Establishing your own grid is important, especially when you're planning to reach well beyond 1,000 days. However, creating an integrated fortress requires careful planning and technical considerations. This chapter will dive into the nitty-gritty of it all—the calculations and the choices that will make your off-grid life easier. It may seem like a lot now, and you won't be doing this until you are well established on your homestead, but laying the groundwork now will have you more prepared for when the time is right.

9.1 Understanding Grid Basics

While we have been over this a bit in our energy solutions chapter, it is time to master the fundamentals of your power source. Understanding the differences between DC and AC systems, electrical concepts, and system sizing are all crucial for building a self-sufficient and secure home grid.

DC vs. AC

Yes, you have a choice! That also means that you are caught in the battle between direct current (DC) and alternating current (AC). Like with anything you do off-grid, there are advantages and disadvantages; however, understanding them a bit better will help you choose the right system for your energy needs.

- **DC:** Think of electricity flowing steadily in one direction, like water coming from your garden hose. That's a direct current. It's simple, requires less equipment, and it's ideal if you're only using smaller appliances and battery storage. However, when you're trying to send that energy across a large homestead, it becomes inefficient because of power loss.

- **AC:** An alternating current is a constantly reversing flow, like a seesaw moving up and down. This dynamic trait allows efficient transmission over long distances, which is why it's standard for household electricity. However, as you saw in the previous chapter, AC systems need additional equipment to convert DC to AC before going to your home.

Knowing the Electric Lingo

There are some very key electrical terms, and knowing them can keep you safe when working on any electrical system.

- **Voltage (V):** Think of this as what pushes the electricity through your system. More voltage allows the transmission of power over longer distances. Most off-grid systems are going to use 12V, 14V, or 48V DC.

- **Amperage (A):** This represents the amount of electricity moving through a conductor, much like water flow in a pipe. More amperage means more power delivery.

- **Wattage (W):** This is the rate at which your appliances consume electricity. You can calculate this by multiplying voltage by amperage. Knowing the wattage of your appliances will help you get the right-sized system for your homestead.

Keeping Your Homestead Safe With Grounding

When installing any energy solutions for your home, proper grounding is crucial for the safety and longevity of your off-grid system. This gives you a safe path for electricity to discharge, which will keep you and all your equipment safe from damage. Your system must have a proper grounding connection to the Earth.

Sizing Your System and Getting the Right Fit

Think about your homestead at full power. You should account for all your home comforts, security systems, and outbuildings. You'll then calculate the total wattage of everything you plan to use. This is what will help determine the right size for your solution (solar panels, turbines, etc.), batteries, and inverter. The right size is needed to meet your energy demands, but you can't forget to factor in seasonal amendments and future expansion.

9.2 Building the Backbone

A well-designed electrical system will form the backbone of your self-reliance. In this section, we will learn the crucial aspects of wiring and distribution, which will empower you to build a safe, efficient, and reliable power network on your own.

Selecting the Right Cables: The Arteries of a Grid

Getting the appropriate cables for your system is crucial, which means you'll need the right wire gauge. A wire gauge, as indicated by numbers like 10 AWG or 14 AWG, determines the amount of current a wire can carry safely. Larger gauges are the ones with lower numbers, and these can handle your higher currents over longer distances with only a small voltage drop. Some online calculators will help you find the right gauge based on your needs and cable length. However, if you are stuck or unsure, consult with an electrician.

The material you choose will also play a huge role. Copper wire has superior conductivity and durability when you compare it to aluminum, which is why you'll find this in most construction projects (especially off-grid). However, aluminum can be ideal for the budget-conscious homesteader who has shorter runs. However, no matter the material, these cables need to be insulated for safety and weather resistance. Don't take any chances.

Minimize Losses and Maximize Power: Making an Efficient System

A smart design keeps your cable lengths shorter, which means less voltage drops. This is how you ensure efficient power delivery throughout your system. Never rush your design. Plan the layout carefully and utilize the shortest possible cable runs while still adhering to safety regulations and avoiding obstructions. If you do have longer distances to cover, consider increasing the wire gauge to make up for the voltage drop. Again, you can always consult with an electrician before you start laying out your wires. It's all about getting the best layout, which means it has to be done right.

Routing Your Conductors

Once your layout is planned, you have to choose the conductor route. You could use underground trenching, which will offer a lot of protection, but it requires careful digging and conduit installation. You'll also need to use a conduit to shield the cables above the ground. This will provide mechanical protection and prevent accidental contact.

You can also use overhead lines, which are sufficient for long distances, but they require support structures and strict safety measures. For both paths, you will have to consult your local regulations. And don't forget to prioritize your safety and the safety of everyone on the homestead.

Components that Are the Building Blocks of Distribution

Now, we have to dive into the components that will manage your power flow.

- **Junction boxes:** These serve as connection points for multiple cables, and they distribute power to different circuits.

- **Fuses and circuit breakers:** These components are there to protect your system from overloads and short circuits. They will automatically interrupt the flow if there is a fault. You will want to choose components that are rated for your system's voltage and amperage.

- **Power hubs:** A power hub can be broken up into three parts: the main panel, the subpanel, and the dedicated circuit:

    ○ The main panel is your central hub. It receives power from your generation source and sends it to various circuits through subpanels.

    ○ Your subpanels can be installed in different areas of your homestead, and they will give you localized power distribution and reduce cable runs.

○ Dedicated circuits are crucial when you have high-demand appliances like water pumps or even your refrigerator. The dedicated circuits ensure that these appliances get adequate power without overloading other circuits.

- **Grounding components:** Again, you need to provide a safe path for stray electricity to discharge. You should have grounding rods driven into the earth to establish a grounding point. This rod is connected to your system with grounding wires. Bonding will connect your metallic components to your grounding system, and this will prevent a voltage buildup. Grounding your homestead is crucial for maintaining safety when off-grid, and it should be implemented in line with local codes and best practices. Again, if you are unsure, get a consultation before moving forward.

Planning and building your off-grid electrical system demands meticulous planning, well-informed choices, and adherence to safety. This section is only meant to give you a foundational understanding of electrical systems. You can certainly DIY this, but that doesn't mean that you shouldn't consult a professional before getting to work. You also have to stick to your local regulations, which will further ensure a safe and reliable off-grid network.

9.3 Connecting the Dots

Now, we have to cover aspects of power generation and management. Navigating this piece of your off-grid power requires your understanding of the interconnected components that make your system tick. So in this section, we will dive into the next set of crucial elements, from converting your energy sources to managing your consumption. These pieces will ensure a smooth and reliable off-grid experience.

DC to AC: The Importance of Inverters

So, you learned before that your energy solution captures direct currents, but usage of DC is severely limited. If you want to set up your homestead, everything is going to demand an alternating current. That's why an inverter is going to be crucial for your homestead. This component steps in and converts the DC electricity from the source or your batteries, and it turns it into usable AC power. The right inverter will be based on your power needs and appliance compatibility. So, let's look at three types of inverters.

- **Pure sine wave:** These are ideal for sensitive electronics (computers, TVs, etc.), and they provide clean and stable power.

- **Modified sine wave:** You will see these and be tempted by their budget-friendly tag. While they can run basic appliances, they might not be the best choice for your sensitive equipment.

- **Hybrid inverters:** These combine inverter and battery charger functionalities, which can simplify your system and even save space. Ensure that the model you choose can meet your basic needs and keep your sensitive equipment safe.

Keep in mind the wattage requirements of all your appliances and make sure the inverter can handle peak loads. Another helpful tip: inverter efficiency will also impact your overall performance, so choose a model with a high conversion rate.

Battery Banks: Your Energy Reservoir

Batteries act as your energy reservoir, holding onto the power that was generated during peak sunlight or windy hours and releasing it when your panels or turbines are at rest. Of course, this means that you will need to have an understanding of different battery types and how to calculate capacity. These are both crucial points, but let's start with the battery types.

- **Lead-acid batteries:** This is an affordable option, but they have a much shorter lifespan and need frequent maintenance.

- **Lithium-ion batteries:** These batteries are of high quality. They have a longer lifespan, higher efficiency, and require less maintenance. The only drawback is the hefty price tag.

- **Deep-cycle batteries:** These batteries are the perfect fit for your homestead. They are designed for repeated charging and discharging.

Now, we need to move on to calculating your battery bank capacity. You'll need to look at your daily energy consumption (kilowatt hours), desired autonomy (days without sunlight or wind), and appliance usage patterns. There are online calculators that will give you accurate results, or you can consult an electrician for this. Keep in mind that proper charging and discharging practices are necessary for battery health. I would suggest investing in a multi-stage charger and avoiding deep discharges when possible.

Charge Controllers for Keeping a Balanced Flow

Charge controllers are the gatekeepers, and they regulate the flow of power between your energy solution and your battery bank. They prevent your batteries from getting overcharged, which can damage them later, and they ensure an efficient energy transfer. When choosing a charge controller, look for features like:

- **Maximum power point tracking (MPPT):** This is for solar systems, but it will optimize the solar panel output, generating the most power.

- **Temperature compensation:** This feature adjusts charging based on the battery's temperature, maintaining optimal performance.

- **Diversion load control:** This will redirect excess power during times of high generation, which is crucial for preventing unnecessary damage to the batteries.

Energy Management Systems for Efficiency

Energy management systems (EMS) are like the brains of your setup, giving you comprehensive monitoring and control. This tracks your power generation, consumption, and battery health, which gives you deeper insight into how your system is performing. The following are some advanced features that allow you to have even more control over this system.

- **Demand forecasting:** This helps predict energy needs, which will then optimize battery usage.

- **Load shedding:** Your system will automatically disconnect non-essential appliances during peak load times to preserve battery power.

- **Remote monitoring:** You can access your system from anywhere to see its status and make adjustments.

Before you get an EMS, you should be aware that they aren't necessary for every system. However, they will significantly enhance efficiency, optimize usage, and give you peace of mind, knowing that you'll have uninterrupted power flowing to your entire homestead.

Automation: Control from Afar

Look into timers and switches to add to your system. Have your lights scheduled to turn on and off automatically. If you have an irrigation system, you can put that on a timer, too. The same goes for your appliances. This is where having a smart hub will really do wonders, as it works in tandem with your home grid.

Safety Measures for Your Off-Grid System

Safety should be a priority with any electrical system. Overcurrent protection devices like fuses and circuit breakers are necessary to prevent overheating and potential fire hazards. These need to be installed at critical points throughout your system to protect your equipment and yourself.

Emergency shutoff switches will give you an immediate way to isolate your system in the event of faults or emergencies. These switches should be easily accessible and clearly labeled.

9.4 Additional Considerations and Optimization

Living off the grid is a huge testament to your commitment to self-reliance. However, the deeper you get into your journey, you'll see that your needs and desires will evolve. That applies to all aspects of your homestead, including your energy system. They all need to evolve, too. But how do you get there? In this last section, we will explore strategies for fortifying and optimizing your energy set up to withstand the long haul. This will ensure that your system is sustainable and gives you reliable power throughout your life off-grid.

Expand the Energy Horizons: Scaling Up for Growth

Your energy consumption will change, which might mean that your generating capacity will need to change as well. This will be particularly true during the first couple of years off-grid. The following is the best way to prepare for your growth:

- **Add solar panels or wind turbines:** If you need to increase power generation, your first route should be to add more panels or turbines if you have the space. This might be the case if you're adding more energy-demanding appliances (you really can't avoid some of them). This would be a great time to look into high-efficiency panels or bifacial models for more energy capture since these are often more expensive than what you'd use in your initial setup.

- **Battery bank expansion:** You might have enough panels, which means your issue might be with your battery setup. Expanding your storage can give you more autonomy, but when you do, make sure your charging system is up to the task. Another issue may be with your initial battery setup. Remember, while it might be more expensive, there are battery options that will meet your needs.

- **Inverter upgrade:** As you expand and evolve, you might run into a wall with your inverter. Consider upgrading to a higher-capacity model, which will accommodate existing and future needs. Remember to look for models that have the advanced features we discussed earlier.

Exploring Alternative Energy Avenues

There are popular choices for energy generation, like solar, wind, and hydro, but you should always be on the hunt for other methods to further your sustainability and resilience.

- **Biogas digesters:** These digesters take organic waste from your homestead and convert it into methane gas, which you can use for cooking, heating, or even electricity generation (through microturbines). This is a great, sustainable waste management solution with the added benefit of additional energy production.

- **Microturbines:** These small turbines use various fuels like biogas, natural gas, or propane. This will give you backup or alternative power sources, especially if you have long periods with limited solar or wind resources. However, you will need to research fuel efficiency and noise levels before you invest.

Powersharing (If in an Off-Grid Community)

If you are fortunate and find an off-grid community, the doors are wide open for all kinds of possibilities for power sharing. You can collaborate with your neighbors on the following:

- **Establish a microgrid:** You and your neighbors can pool your resources and create a shared grid. This will enable power exchange and backup support, which will virtually eliminate any dark periods. It will require some significant planning and investment, but there are huge long-term benefits, like increased reliability and cost-effectiveness. You even open up the option of grants or community funding for microgrid development.

- **Peer-to-peer sharing:** You can utilize battery storage systems to exchange excess power directly within the community. This will enhance overall efficiency and resource utilization.

The thing to remember when in an off-grid community is that there needs to be clear communication, defined agreements, and careful planning with your system design. This will likely mean bringing in legal professionals to ensure agreements are adhered to and that they comply with local regulations.

Beyond Batteries: Alternative Storage Solutions

Batteries are the most common storage option at the moment, but there are emerging technologies that you might want to consider or keep in mind when you expand your system.

- **Flywheel energy storage:** These systems will store energy in spinning rotors, which gives you a fast response time and high discharge rates. That makes these great for short-term fluctuations. I would leave this as a maybe for now, though, and the initial cost is steep.

- **Hydrogen fuel cells:** The infrastructure for this storage method and distribution is still being developed, but it's something to look out for. These cells convert hydrogen into electricity through an electrochemical process, giving you clean and efficient energy storage. This is especially true when paired with renewable hydrogen production through electrolysis.

Again, these methods are still in the early stages, so they are expensive or simply not feasible for just anyone to use. However, you should keep an eye on these technologies as you expand your homestead.

Responsible Disposal

When you are living off-grid, you are already practicing responsible waste management. This is going to include batteries, inverters, and other electrical components, as they all have specific disposal requirements. Follow proper protocols when recycling or disposing to scale down on the environmental impact. Research recycling facilities in your area or contact organizations that specialize in electronic waste management.

***

Your off-grid experience will be a continuous learning process. Make sure you regularly monitor your system's performance, keep track of your energy consumption data, and never be afraid to adapt your strategies as needed. Stay informed about technological advancements or attend workshops to learn more. Again, while you can DIY your homestead's power, getting consultations and guidance from professionals like contractors and electricians can prevent you from making a costly mistake.

Everything we have learned through nine chapters is everything you will need to make your first 1,000 days off-grid a success. But we still have one more thing to cover, and that's even more DIY projects that will benefit your homestead.

Chapter 10:

Beyond the Basics—More DIY Projects

This chapter is dedicated to just a few more DIY projects that can help any homesteader in their off-grid journey. As always, stay safe when working on any of these projects. If you are unsure about anything, don't hesitate to do some further research to find diagrams or more simplistic designs.

Biogas Digester

This DIY project is something that was mentioned toward the end of the last chapter. This system harnesses the power of nature by turning your organic waste into methane gas, which means you have a renewable fuel source for cooking, heating, or even lighting the home.

Materials

- The digester tank

    - large plastic container (should be airtight and leak-proof).

    - metal drum (optional) for durability, but might need welding and other modifications.

    - concrete blocks (optional) for support and stability when using metal drums.

- Gas collection system

    - gas outlet pipe with a valve: Ensure that you are using a material compatible with biogas, like copper or stainless steel.

    - gasometer: This is optional, but this will regulate gas pressure for a safer, more efficient system.

    - Biogas hoses and connectors: Choose hoses that are made for biogas applications. This prevents leaks later on.

- Feed inlet and outlet

    ○ pipes or hoses with valves: This will control the feeding of organic materials and the removal of the digested slurry.

    ○ mesh screen: This is another optional but ideal component, as it will prevent larger materials from clogging your system.

- Mixing system

    ○ paddle or pump: The mixing system is optional, but regular stirring will promote even digestion and will prevent scum from forming.

Instructions

1. You'll start by preparing your digester tank. You will want to test your container a few times to make sure that it is airtight and leak-proof. If you use a metal drum, remember that you might need to make modifications. If you want to use the slurry later for fertilizer, then also make sure that the material is food-grade.

2. Now you'll install the feed inlet and outlet. You can do this by creating openings on opposite sides of the tank. One side will be for adding organ waste (feed), and the other side will be for the removal of the slurry. Install the valves and optional mesh screens at this time. Make sure that when you create the openings, you consider the angle and size to ensure optimal flow.

3. The tank is now ready for the gas collection system. Install a gas outlet pipe with a valve at the top of the tank. Hook this pipe up to your gasometer and then to the biogas hoses and connectors. Remember to choose materials and their sizes based on your expected production and intended uses.

4. Now you can prepare your organic waste. Chop and mix food scraps, manure, and other organic materials. Just like with your compost, you will want to avoid meat, bones, and oily substances because they will hinder the digestion process, reducing your gas production. You also want to consider the moisture content and particle size.

5. Once your feed is ready, add this and water to the digester tank. Think 70–30 here, with 70% water and 30% feed. Like with hot composting, you'll want to maintain a good temperature. In this instance, between 65 °F and 95 °F is a great range for optimal activity and gas

production. Depending on your climate, you might consider options for insulation to maintain that temperature.

6. If you choose to use a mixing system, you'll want to regularly stir the contents.

7. After several weeks, you will have methane gas that needs to be collected from the gas outlet. When doing this, you should use appropriate safety precautions. This means making sure all appliances and connections are in working order with no leaks. Biogas is a flammable material, which is why proper handling and regular harvesting (venting) are crucial.

Spring Box

This project is a simple structure, yet it safeguards your natural spring water while making it easily accessible. Before you begin this project, ensure that you know the local regulations that involve your water access.

Materials

- Foundation

    - crushed gravel or small rocks

    - leveling sand

    - landscape fabric (optional)

- Box walls

    - wood planks or logs: It might be more expensive, but you should choose naturally rot-resistant woods like cedar or redwood.

    - waterproof lining: This can be a pond liner or another heavy-duty plastic.

    - concrete blocks: This is optional, but they will add a bit more stability.

- Spring outlet

    - pipe and valve: Choose materials that are compatible with drinking water.

    ○   diverter: This is also optional but allows you to direct water flow for multiple uses.

- Additional items

    ○   grave or stones (for drainage)

    ○   rocks or plants (optional for decoration)

Instructions

1. You want to start this project by carefully selecting a location. Identify the source and make sure there is proper drainage around it, and then mark the area for your spring box.

2. After you have the area marked, you will prepare the foundation by digging a shallow pit slightly larger than the dimensions of your box. You'll fill this with gravel (or small rocks) for drainage, followed by a layer of leveling sand. This will create a stable base. To prevent soil erosion, I would consider making your first layer the landscape fabric, and then the gravel.

3. Now you get to make the box walls. There are three ways you can go about this, so consider it before gathering your materials.

    A. **Wood:** You'll assemble your planks or logs into a box shape. You will need to have some mastery in woodworking because you want to ensure that there are tight seams and no gaps. This prevents water leaks. After the box is made, you'll apply waterproof sealant to the interior surface.

    B. **Concrete blocks:** This is a bit simpler, as you'll stack the concrete blocks to make your box. Apply mortar to secure them, and then apply waterproof sealant on the interior surface.

    C. **Hybrid:** For this, you'll make the outer wall with concrete blocks and mortar. Then, on the inside, you will attach wood planks to give the box a finished look. You'll still apply sealant on the inside.

4. After everything has set, you'll line the interior. Line the entire box interior with the pond liner, and make sure that it extends beyond the top edge. You'll seal all seams and edges to prevent leaks and any potential contamination.

5. Now you'll need to create the outlet. Install a pipe and a valve through the box wall at the point where you want to get water. This is when you'll use your diverter pipe, too.

6. For the finishing touches, fill the bottom of the box with gravel or stones. You can also cover the exposed liner around the edges with your decorative elements. Also, this is the time to make sure that water is flowing freely and that there is adequate drainage.

Solar Water Heater

Harnessing the sun's energy to heat your water is an eco-friendly way to give your homestead a sustainable alternative to your conventional water heater. This is ideal for times when there is low power or you need to conserve energy. While there are various DIY solar water heaters, this set of instructions will focus on an overview of the flat-panel system, which is great for moderate hot water needs.

Materials

- Collector Box

  - plywood sheets: The size will be based on the desired collection area.

  - black paint: Opt for absorbent and high-heat-resistant paint.

  - insulation boards (Rockwool or foam)

  - glazing material: I would recommend using tempered glass or polycarbonate sheets.

  - wood screws and weather-resistant sealant

- Circulation system

  - copper tubing: Select a diameter that is appropriate for water flow

  - header pipes: These will connect the individual tubes.

  - hose clamps and pipe fittings

  - water pump: Look for one that is suitable for solar use.

  - You can also get a flow meter, temperature sensors, and a controller, but these items are optional.

- Storage tank

  - insulated water tank: The size will be based on the daily hot water needs of the homestead.

  - expansion tank: This will help you accommodate any volume changes.

  - valves and plumbing connectors

- Additional materials

  - wood for frame construction (optional)

  - legs or a stand to elevate the collector

Instructions

1. Build the collector box.

   A. Cut your plywood sheets to your desired dimensions. These sheets should also fit your chosen glazing material. Using your wood and screws, you'll create a sturdy box.

   B. Paint the inside of the box black for heat absorption. To keep heat loss to a minimum, apply the insulation to the back and sides of the box. Drill out a hole that will be the inlet and outlet for the tubing.

   C. Save this step until after the copper tubing inside the box. The glazing material will be put over the open end of the box, and you'll secure it using your weather-resistant sealant. This seal should be watertight.

2. Assemble the copper tubing.

   A. Design the tubing layout to go inside the collector box. You want to cover as much of the box as you can without compromising water flow.

   B. Bend the tubing into snake-like loops that follow your design. You'll connect the individual tubes using header pipes.

   C. Secure the tubing inside the box with straps or brackets, and make sure the inlet and outlet pipes are out of the box. Seal the area around the tubing and place the glazing material over it.

3. Connect the circulation system

   A. Attach the inlet and outlet pipes to your water pump. If you opted to use a flow meter and temperature sensors for system monitoring, install those components here.

   B. Connect the pump outlet to the bottom of your insulated water tank. If using an expansion tank, install this at the highest point of the system.

   C. Run separate cold water inlet and hot water outlet pipes into your existing system. Make sure you clearly label and use valves for isolation purposes.

   D. If using a controller, it should be programmed to operate the pump based on your desired water temperature and sunlight availability.

4. Mount and position the collector

   A. Like with any solar system, choose a south-facing location that gets maximum sunlight exposure. You should also ensure that the location can support the weight of the box.

   B. You can then elevate the collector on a frame or stand to get optimal water flow and drainage. You can adjust the angle of the box based on the sun's positioning to get maximum exposure.

5. Test and monitor

   A. After all your connections are closed up, fill the system with water. Make sure there are no leaks before turning your pump on. You will then test the pump operation to ensure that you're getting proper water flow through the collector.

   B. Monitor the water temperature in the tank for a few days to assess its performance. You can fine-tune the pump operation or adjust the box's tilt if needed.

DIY Gray Water System

When you are off-grid, conserving your water resources and keeping environmental impact low are key principles of building a sustainable life. One issue that some homesteaders have is what to do with the water that drains out of their showers, sinks, or washing machines. A greywater system is just what you need, allowing you to reuse that water for non-potable applications like garden irrigation, toilet flushing, and other uses. This saves precious gallons of potable water to be used in the home.

Materials

- **Diverter valves:** You will want to choose valves that are compatible with your home plumbing system, particularly the greywater sources. This will save a lot of work and materials later. The size of the valve should be based on your anticipated water flow and building codes. You will also need to get gray water-safe hoses to connect to the valves.

- **Filters:** There are three types to choose from.

  - **Sand filter:** This will give you basic filtration, which means large particles and debris. Of course, the size should match your anticipated greywater volume and how much treatment you want that water to have.

  - **Biological filter:** This filter uses bacteria to break down organic matter, which gives you better-treated water. You can get a pre-built filter, but there are also DIY solutions for this as well.

  - **Membrane filter:** This will be your advanced filter, and it will remove bacteria and other contaminants. With those advancements, though, will come regular maintenance, which could be costly.

- **Storage tank:** You will need to get a dedicated storage tank that is specifically for greywater. You will need to weigh factors like material (plastic or concrete), size (based on expected usage), and location (for easy access).

- **Distribution system:** Use separate pipes and valves for distribution. Make sure the materials are suitable for non-potable water, and they must be clearly labeled to avoid confusion with your potable water systems.

Instructions

1. The first thing is to identify your gray water sources. Your best options would be your shower, sinks (except kitchen sinks because of grease), and washing machines. Basically, any sources that have chemicals, grease, or high fecal matter content are out of the question, as they will need more complex treatment or are unsuitable for reuse.

2. You'll then install your diverter valves; so locate the drain lines from your source. Install the valves near these lines to redirect the used water flow toward your treatment system. Make sure that you have proper connections and leak-proof seals.

3. Now you'll build the treatment system. This is going to vary based on the filter you choose, financial considerations, and building codes.

   A. **Sand:** For this filter, you'll construct a concrete or plastic tank filled with sand layers of various sizes. The water will flow through those layers, removing debris and particles.

   B. **Biological:** No matter if you buy or build your own filter, you will need to consider things like oxygenation and flow rate.

   C. **Membrane:** Choose a pre-built system with membrane types and capacities based on your needs. All you'll need to do is ensure proper installation and maintenance.

4. After you have your filter system in place, you'll connect the diverter valves to your filter. You'll want to ensure that there is proper flow direction, and you'll want to pay close attention to the volume of gray water. Your flow might require a larger filter or a combination of filters.

5. Once you're sure that the filtration system meets your needs, you can install the storage tank. Remember to have this in an area with good drainage and easy access for maintenance. Connect the filtered water to the tank and put in an overflow pipe to prevent overfilling.

6. Now you can run separate pipes from the tank to your reuse points. One of the ideal places to run it in your garden during the growing season is for irrigation purposes. You can also have pipes directing the water to other areas that don't need potable water.

Raised Garden Bed

Sometimes, we have to work with what we have in terms of soil, but that shouldn't stop you from creating a thriving haven for vegetables or herbs. Raised beds allow better drainage, soil control, and easier access, which can save a lot of your energy when tending to your plants. With the numerous advantages, it should be no surprise that you'll find a couple of raised beds on most homesteads and in many backyard gardens.

Materials

- **Lumber:** Again, weather-resistant wood like cedar or redwood works great, or you can opt for pressure-treated lumber. The thickness will depend on your durability needs and bed size, but the best route is typically with 2x4s.

- **Fasteners:** You should use galvanized or deck screws. These are rust-resistant and will make the bed sturdy.

- **Landscaping fabric:** This will prevent weed growth and maintain soil moisture retention.

- **Soil:** You can either purchase high-quality soil from a garden center, or you can make your own with a blend of compost, topsoil, and organic matter.

- **Tools:** You'll need a measuring tape, saw, drill, screwdriver, level, shovel, and gloves.

- **Optional materials:** You can get stakes, edging material, or decorative elements, but they aren't necessary.

Instructions

1. Your first step comes in the form of planning. Two areas are going to factor into those plans for raised garden beds.

   A. **Size and location:** You'll need to choose a spot on your land that gets at least 6-to-8 hours of sunlight daily. The size of the bed, though, will be based on how much space you have and what you want to plant in the bed. It can be as big as you want, but make sure you can reach all areas of the bed easily for maintenance.

   B. **Bed height:** Decide on the optimal height for the bed. This should consider your comfort and planting preferences. Taller beds can provide better drainage and easier access, but lower beds are better for smaller plants or aesthetics. You can also use various heights, as long as you're comfortable.

2. Once you have a plan, it's time to prepare the site. Start by clearing the area. Remove any existing vegetation, rocks, and other debris from the site. From there, using your chosen method, outline the dimensions of the bed. The site should then be leveled. Leveling is crucial to prevent water pooling and soil erosion.

3. Now it's time to build the frame. Cut your boards to the lengths that were determined by your plans. Lay them flat and pre-drill holes to prevent them from splitting as you assemble the frame. Join your boards at the corners, making sure everything is squared and level. Once everything looks as it should, securely fasten the connections with the screws. If your bed needs multiple walls, just repeat step three and attach them to the existing frame.

4. With the frame now in place, lay down the landscaping fabric. This should cover the entire base and extend slightly up the walls of the frame. You can then staple it or use landscape pins to hold it in place.

5. It's finally time to add the soil. Whether you made your own or purchased it, the soil should be loose and well-draining. Don't just dump the soil in, though. You'll want to do this gradually until you've filled three-quarters of the bed. This opens up space for settling and mulching later on. Lightly tamp down on your soil to make sure there are no air pockets.

6. You can finish the bed off with some additional touches. You can use stones, bricks, tin, or other materials to make a decorative border. If you're in a drier climate, you can apply a layer of organic mulch around the plants to retain moisture. Mulch will also help suppress weeds and regulate soil temperatures. And, of course, plant the things you planned for. As long as you follow instructions for their care, your plants will thrive in this bed.

Additional Tips

- If the base of your bed is solid, you can drill small drainage holes near the bottom of your frame. This will prevent waterlogging, which can damage your plants.

- You should start small. Spend a season with a small bed, growing herbs and some other vegetables. From there, you can expand your raised bed garden to host even more plant life.

2-DIY Rocket Stove Ideas

Rocket stoves have become staples for off-grid homesteaders, preppers, and anyone with a knack for DIY. They burn wood or other biofuels in an incredible, efficient manner, which makes them perfect for camping, survival situations, or just having a cookout on the homestead. Rocket stoves are a project that anyone can do, and for this section, we will work through two unique designs that you can implement now or on the homestead.

Design 1: The 55-gallon Drum

This design is durable and offers ample cooking space, which makes it perfect for larger groups or long outings.

Materials

- 55-gallon drum

- 20 firebricks

- 12in. x 24in. metal sheet

- 2in. x 2in. angle iron (24in. long)

- 12in. x 12in. metal plate

- drill

- saw or grinder

- screwdriver

- bolts and nuts

Instructions

1. The first thing you'll do is cut the drum. Remove the top of the drum with your saw or grinder. Then, you'll mark and cut a 2 x 4 in. rectangular hole near the bottom. This hole will allow air to get in.

2. You'll then line the drum with firebricks. Arrange them in a circle inside the drum, but make sure to leave at least a 1-inch gap between the bricks and the wall of the drum.

3. To make the air channel, bend the angle iron until it makes an "L" shape. The longer end of the "L" will be attached to the drum below the air inlet hole. Secure it with nuts and bolts. You'll then cut a rectangular shape out of the sheet metal that matches the size of the inlet. Attach the sheet metal piece to the angle iron to make the air channel.

4. Cut a circle near the top of the drum, around 4 inches in diameter. Use the rest of your sheet metal to make a chimney with a cap. Secure that to the drum with nuts and bolts.

5. Take the metal plate and drill holes in it. Make sure the holes are evenly spaced, as this will become your cooking surface.

The Upcycled Can

This is a quick and easy design, which is perfect when you're in a pinch and away from home.

Materials

- two tin cans (different sizes are ideal)

- 4 x 4 in. metal mesh

- can opener

- hammer

- drill

- pot stand

Instructions

1. Both cans should be empty and washed out. If not, use the can opener to open one end of both cans. You're then going to cut a 2 x 3 in. rectangular hole near the bottom of the larger can, which will be your air inlet.

2. Now you'll make the fuel chamber by nesting the smaller can inside the larger one. There needs to be a gap near the bottom for airflow. This can be secured with wire, or you can use rivets.

3. Now, you can add the mesh and the stand. Place the mesh over the air-inlet hole. This will keep embers from getting out. The pot stand will go on top of the larger can.

4. Your fuel will go in the smaller can, which will give you efficient flames for very little fuel.

Rocket stoves don't stop there either. Some designs use clay, some use cinder blocks, and even some more complex designs require some know-how in the metalworking department. You might even have an efficient system of your own. When it comes to a rocket stove, there is plenty of room to get creative.

***

And that brings us to our bittersweet conclusion. In a few short chapters, you have been introduced to some of the most unique off-grid projects that just need some basic materials and an innovative spirit. Your home will hum with self-reliance as a testament to your ingenuity and grit. Although we have reached the end of our projects, think of them as just one page of the giant book on DIY projects.

Every corner of the world around you have untapped potential. The dusty toolbox and "junk" in your garage hold inventions that we haven't seen yet. Those old cans can become an expedient rocket stove, and let's not forget everything in the attic that can be repurposed. The real key to all this is in you. You just have to unlock your creativity and let the inventor in you run wild.

So don't pack your tools up just yet. Let the projects we've learned be your springboard. With these, the ones you've yet to learn, and all the off-grid advice we have covered, you are ready to tackle anything, anywhere, at any time.

LEAVE A 1 CLICK REVIEW!

Customer Reviews

★★★★★ 2
5.0 out of 5 stars ▾

5 star	▓▓▓▓	100%
4 star		0%
3 star		0%
2 star		0%
1 star		0%

See all verified purchase reviews ›

Share your thoughts with other customers

Write a customer review ⬅

Since self-publishers really depend on readers' reviews, I would be incredibly grateful if you could take 60 seconds to write a brief review on Amazon, even if just a few sentences, so that people around the world may discover it and find it helpful, as well!! Please click on the link below to leave a review. Thank you!

Conclusion

This is more of a beginning than an ending. You are now ready to tackle 1,000 days off-grid. This journey is going to be a testament to your grit, resourcefulness, and resilient spirit. This journey was about so much more than just surviving; it was about thriving, connecting with your surroundings, and making your future one that is built on your terms. As you look toward that future, remember the lessons you learned and the challenges that you have yet to overcome.

This book gave you a glimpse into not just what you can do off-grid but all the realities that off-grid living presents. You've learned how to harness your resources for energy production, crop tending, hunting, and gathering, livestock raising, and security. While the lessons and the DIY projects provided practical skills, the things you should have taken away lie deeper. Let's look at those.

- **Know yourself:** Before taking that step to an off-grid life, know what motivates you, what your expectations are, and what your vulnerabilities are. Off-grid living means that you have to be self-reliant, resilient, and willing to adapt to anything.

- **Planning pays off:** Never underestimate the power of careful planning. From drafting and perfecting a budget and researching regulations to choosing the right equipment and designing your homestead, preparation is the fastest way to success.

- **Never stop learning:** Off-grid living comes with a continuous learning curve. Every day is going to present new challenges, and every solution will give you a wealth of experience. Stay curious; even if you have something established, there might be another solution that will benefit you more. Get knowledge from various sources and connect with your fellow off-gridders for shared support and learning.

- **Minimize impact:** Practice responsible resource management, find sustainable solutions, and leave as little trace as you can.

- **Community is key:** No homesteader should exist alone. You should establish connections with any neighboring homesteaders that can lead to resource sharing. Or you can forge bonds with homesteaders you meet online, which establishes a much-needed support network.

- **Just embrace the unknown:** Off-grid living isn't easy. This is a journey of challenges, detours, setbacks, and moments of pure joy. Embrace the unpredictable, celebrate your wins, and learn from the losses. It's all of these experiences combined that shape the off-grid life you dream of.

This book was not a definitive guide. This was more like a companion on your journey. That means you should refer to it often. Revisit the information as your needs evolve, and share this with others who are considering a step into an off-grid lifestyle.

Remember, the road may be demanding, but the rewards are immense. Start roaring today, gather your courage, and take your first steps into a self-sufficient tomorrow. The off-grid awaits, and you're ready to conquer it.

References

Ana. (2022, March 23). *Our essential guide to foraging: Tips, books, guidelines.* Foraging Guru. https://foragingguru.com/foraging-guide/

Babington-Stitt, T. (2022, August 5). *Garden zoning – 15 ways to divide your yard and maximize outdoor space.* Homesandgardens.com. https://www.homesandgardens.com/advice/garden-zoning

Bailey, E. (2023, September 3). *How to make homemade solar water heater: A step-by-step guide.* Solar Panel Installation, Mounting, Settings, and Repair. https://solvoltaics.com/how-to-make-homemade-solar-water-heater/#google_vignette

Benny, A. (2014, January 25). *Basics of caring for livestock - different animals certain requirements.* Rural Living Gardening | Hydroponics | Generators. https://rurallivingtoday.com/livestock/basics-caring-livestock/

Boeckmann, C. (2024, January 10). *What are plant hardiness zones?* Almanac.com. https://www.almanac.com/what-are-plant-hardiness-zones

Bradshaw, A. (2019, January 19). *Home security for the off-grid (or any) homestead.* My Homestead Life. https://myhomesteadlife.com/home-security/

Brotak, E. (2020, December 1). Microclimates in gardens: Their causes and your opportunities. *Horticulture.* https://www.hortmag.com/gardens/microclimates-in-gardens

Burnley, R. (2021, October 24). *14 questions & answers about getting A fishing license.* Kayak Angler. https://kayakanglermag.com/gear/fishing-gear-accessories/fishing-license/

Byrd, J. (2023a, October 17). *Which water filter removes the most contaminants?* Https://Waterfilterguru.com/. https://waterfilterguru.com/which-water-filter-removes-the-most-contaminants/

Byrd, J. (2023b, November 21). *The 2024 ultimate guide to water purification.* Waterfilterguru. Https://Waterfilterguru.com/. https://waterfilterguru.com/water-purification/

Cal, R. (2021, April 3). *The ultimate guide to off-grid composting basics.* Rustic Skills. https://rusticskills.com/homestead-gardening/ultimate-guide-composting-basics/

Campbell, B. (2023, October 17). *18 methods for off-grid water filtration & purification*. Waterfilterguru. https://waterfilterguru.com/off-grid-water-filtration-purification/

Chas. (2023, April 14). *Why DIY? Exploring the benefits and satisfaction of do-it-yourself projects*. Chas' Crazy Creations. https://chascrazycreations.com/diy/

Clayton, D. (2021, June 17). *Types of wind turbines: HAWT, VAWT and more explained*. Energyfollower. https://energyfollower.com/types-of-wind-turbines/

Collings, P. (2022, September 18). *Checklist for living off the grid [your ultimate guide]*. Outdoor Happens. https://www.outdoorhappens.com/checklist-for-living-off-the-grid/

Coogan, K. (2023, February 9). *How to build a biogas digester*. Motherearthnews. https://www.motherearthnews.com/diy/biogas-digester-construction-zm0z23zkgar/

Cork, J. (2023, December 19). *How to install an off-grid solar power system*. Family Handyman. https://www.familyhandyman.com/project/off-grid-solar-power-system/

Crank, R. (2019, March 13). *Determining the best renewable energy source for your homestead*. Countryside. https://www.iamcountryside.com/homesteading/best-renewable-energy-source-for-your-homestead/

Crosbie, D. (2022, January 9). *How to build A DIY greywater system (complete guide)*. Climatebiz. https://climatebiz.com/diy-greywater-system/

Davidson, J. (2021, January 6). *Off grid water systems: 4 proven ways to bring water to your homestead*. Tiny Living Life. https://tinylivinglife.com/learn-how-to-build-off-grid-water-system/

Dvorak, T. (2024, January 23). *How to build raised garden beds*. Family Handyman. https://www.familyhandyman.com/project/how-to-build-raised-garden-beds/

E, S. (2022, December 23). *Off-Grid homesteading VS. prepping*. Self Sufficient Projects - DIY Projects to Become Self Sufficient. https://selfsufficientprojects.com/off-grid-homesteading-vs-prepping/

Ellison, M. (2017, June 23). *Summer livestock shelter considerations*. Agproud. https://www.agproud.com/articles/49309-summer-livestock-shelter-considerations

Emily. (2022, August 27). *9 things to know about off-grid homesteading*. Accidental Hippies. https://www.accidentalhippies.com/off-grid-homesteading-things-to-know/

Fowler, J. (2023a, March 10). *The art of wilderness foraging: A beginner's guide*. Practical Off-Grid Living. https://www.practicaloffgridliving.com/the-art-of-wilderness-foraging-a-beginners-guide/

Fowler, J. (2023b, June 3). *Safe and sustainable foraging: A how-to guide*. Practical Off-Grid Living. https://www.practicaloffgridliving.com/safe-and-sustainable-foraging-a-how-to-guide/

Fowler, J. (2023c, August 15). *Wild edibles: A beginner's guide to foraging*. Practical Off-Grid Living. https://www.practicaloffgridliving.com/wild-edibles-a-beginners-guide-to-foraging/

Gallagher, K. (2021, July 16). *Going off-grid with solar panels: Everything you need to know*. Treehugger. https://www.treehugger.com/going-off-grid-with-solar-panels-5190819

Gibson, E. (2022, December 18). *How do you make a natural fishing line?* Trickyfish. https://trickyfish.net/how-do-you-make-a-natural-fishing-line/

Grant, B. (2021, May 16). *StackPath*. Gardeningknowhow. https://www.gardeningknowhow.com/garden-how-to/propagation/seeds/seedlings-after-germination.htm

Harbour, S. (2022, March 18). *Off grid living preparation: 10 things to do now*. An off Grid Life. https://www.anoffgridlife.com/off-grid-living-preparation/

Harler, M. (n.d.). *20 misconceptions about living off the grid debunked*. MSN. https://www.msn.com/en-us/news/technology/20-misconceptions-about-living-off-the-grid-debunked/ss-BB1iOQBO#image=3

Heggie, J. (2021, February 8). *Day zero: Where next?* National Geographic. https://www.nationalgeographic.com/science/article/partner-content-south-africa-danger-of-running-out-of-water

Holpuch, A. (2018, August 8). *The story of a recovery: How hurricane maria boosted small farms*. The Guardian. https://www.theguardian.com/world/2018/aug/07/the-story-of-a-recovery-how-hurricane-maria-boosted-small-farms

How does a solar panel work: Step by step - qcells north america. (2024, February 22). Us.qcells. https://us.qcells.com/blog/how-does-a-solar-panel-work-step-by-step/

How to set up a complete off the grid living system using solar and wind turbines. (2016, September 24). Practical Survivalist. https://practicalsurvivalist.com/offgrid-wind-and-solar-power-backup-system/

Hunt, S. (2019, May 21). *Build spring water collection system with a spring box*. Practical Preppers. https://practicalpreppers.com/how-to-develop-a-spring-with-a-spring-box/

Huynh, T. (2021a, August 16). *8 surprising benefits of living off the grid*. Emoffgrid. https://emoffgrid.com/benefits-of-living-off-the-grid/

Huynh, T. (2021b, December 21). *10 best homestead animals (chicken, rabbits, ducks and mores)*. Emoffgrid. https://emoffgrid.com/homestead-animals/

Jfisher46. (2023, January 11). *Off-Grid living 101: Understanding the benefits, challenges, and realities of living self-sufficient*. Riverbed Ranch. https://www.riverbed-ranch.com/post/off-grid-living-101-understanding-the-benefits-challenges-and-realities-of-living-self-sufficient

Jones, O. (2021, April 30). *10 DIY horse hay feeders you can build today (with pictures)*. Pet Keen. https://petkeen.com/diy-horse-hay-feeders/

julinasmall. (2021, November 8). *9 essential off grid skills every homesteader should know*. Little Dog Ranch. https://www.littledogranch.com/post/9-essential-off-grid-skills-every-homesteader-should-know

Kate. (2022, January 13). *Best off-grid home protection systems in 2023*. Inthralld. https://inthralld.com/best-off-grid-home-protection-systems/

Kylene. (2019, April 10). *The prepper's guide to securing your home*. Theprovidentprepper.org. https://theprovidentprepper.org/the-preppers-guide-to-securing-your-home/

Lampert, E. (2023, August 30). *Off-Grid living: 7 common power systems and how to use them safely*. Green Building. https://greenbuildingcanada.ca/off-grid-living/

Laurens, E. (n.d.). *How to build a homemade water turbine generator*. HomeSteady. https://homesteady.com/12201924/how-to-build-a-homemade-water-turbine-generator

Lee, S. (2022, November 6). *7 simple steps to build A gravity fed watering system*. Bournebright. https://bournebright.com/gravity-fed-watering-system/

Lucy. (2022, January 12). *15 DIY minnow trap projects: How to make A minnow trap*. Little Lovelies. https://www.littleloveliesbyallison.com/diy-minnow-trap-projects/

Maring, J., & Sutrich, N. (2023, January 3). *How to turn an old Android phone into a security camera*. Android Central. https://www.androidcentral.com/how-turn-old-android-phone-security-camera

Meissner, N. (n.d.). *Off grid hydro power 101*. Off Grid Boot Camp. https://www.offgridbootcamp.com/off-grid-hydro-power-101/

Nicolas, N. (2023a, August 7). *Nature's bounty: Hunting and trapping in off-grid living*. OffGridMentor. https://offgridmentor.com/off-grid-hunting-and-trapping/

Nicolas, N. (2023b, August 10). *Go with the flow: Hydro power for off-grid living*. OffGridMentor. https://offgridmentor.com/hydro-power-off-grid/

Nicolas, N. (2023c, August 31). *Hook, line, and sinker: Fishing in off-grid living*. OffGridMentor. https://offgridmentor.com/fishing-in-off-grid-living/

Nubie, S. (2022, February 28). *6 DIY trip wire alarms for off-grid security*. Urban Survival Site. https://urbansurvivalsite.com/diy-trip-wire-alarms/

Off Grid World Team. (2018, February 18). *Off grid security: How to keep your home safe while living off the grid*. Offgridworld.com. https://offgridworld.com/off-grid-security-keep-home-safe-living-off-grid/

Off-Grid water sources and options. (2023, July 31). Off-Grid Rebel. https://offgridrebel.com/off-grid-water-sources-and-options/

Open Green Energy. (n.d.). *DIY solar motion sensor security light*. Instructables. https://www.instructables.com/DIY-Solar-Motion-Sensor-LED-Light/

Pavlis, R. (2022, September 23). *Decoding planting zones: USA hardiness map*. Grit. https://www.grit.com/farm-and-garden/planting-zones-usa-zm0z22ndzawar/

Perry, C. (2024, January 6). *Off grid security cameras: 2024 home defense guide*. Persurvive. https://persurvive.com/off-grid-living/off-grid-security-cameras/

Pleasant, B. (2011, March 2). *40 gardening tips to maximize your harvest*. Motherearthnews. https://www.motherearthnews.com/organic-gardening/gardening-tips-zm0z11zsto/

Poindexter, J. (2016, October 26). *21 DIY rocket stove plans to cook efficiently with wood*. Morning Chores. https://morningchores.com/rocket-stove-plans/

Poindexter, J. (2019, July 4). *23 awesome DIY rainwater harvesting systems you can build at home*. MorningChores. https://morningchores.com/rainwater-harvesting/

Policies and regulations for private sector renewable energy mini-grids. (2016, September). IRENA. https://www.irena.org/publications/2016/Sep/Policies-and-regulations-for-private-sector-renewable-energy-mini-grids

Rinella, S. (2018, August 14). *The hows and whys of hunting licenses and regulations*. Themeateater. https://www.themeateater.com/hunt/small-game/the-hows-and-whys-of-hunting-licenses-and-regulations

Savannah. (2022, November 5). *How to start an off grid homestead: 9 things you need to know*. Survival World. https://www.survivalworld.com/off-grid/how-to-start-an-off-grid-homestead/

Schwartz, D. B. (2018, August 9). *11 living fences that look better than chain link*. Bob Vila. https://www.bobvila.com/slideshow/11-living-fences-that-look-better-than-chain-link-47520

Shallcross, L. (2023, October 17). *DIY whole house water filter system (2024 ultimate guide)*. Waterfilterguru. https://waterfilterguru.com/diy-whole-house-water-filter-system/

Small-Scale livestock production. (2011, September). ATTRA. https://attra.ncat.org/publication/small-scale-livestock-production/

Taylor, G. (2017, July 21). *How to Make a Concrete Catch Basin*. HomeSteady. https://homesteady.com/12740954/how-to-make-a-concrete-catch-basin

Texas Real Food. (n.d.). *Off-Grid water management for homesteads: Essential techniques for collecting and storing water*. Discover Real Food in Texas. https://discover.texasrealfood.com/off-grid-living/off-grid-water-management

The definitive guide to chicken tractors and 13 free DIY plans. (2021, July 9). The Happy Chicken Coop. https://www.thehappychickencoop.com/chicken-tractor/

Tips for successful gardening in off grid living. (2023, September 29). Off Grid Harmony. https://offgridharmony.com/tips-for-successful-gardening-in-off-grid-living/

29 DIY animal trap plans you can make yourself. (2021, December 13). The Daily Gardener. https://www.thedailygardener.com/how-to-build-a-animal-trap

Understanding local regulations before building off-grid home. (2023, January 10). Off Grid Living. https://offgridliving.net/understanding-local-regulations-before-building-off-grid-home/

USDA plant hardiness zone map. (2020). United States Department of Agriculture. https://planthardiness.ars.usda.gov/

Van De Walle, G. (2022, December 10). *First in, first out (FIFO): What food handlers must know*. FoodSafePal. https://foodsafepal.com/first-in-first-out-fifo/

Vartan, S. (2021, October 8). *How to compost at home.* Treehugger. https://www.treehugger.com/how-to-compost-at-home-5119210

Vernazza, L. (2023, April 13). *Building an off-grid house: What you need to know.* The Plan Collection. https://www.theplancollection.com/blog/building-an-off-grid-house

Vuković, D. (2020, May 1). *The off-grid laws of every state in America.* Primal Survivor. https://www.primalsurvivor.net/living-off-grid-legal/

W, E. (2022, January 28). *How to choose the "perfect" location for your off grid homestead or community.* Offgridworld. https://offgridworld.com/how-to-choose-the-perfect-location-for-your-off-grid-homestead-or-community/

Whatisthisidontunderst. (n.d.). *Biogas digester.* Instructables. https://www.instructables.com/Biogas-Digester/

Why you should have a shelter for your cattle. (2021, January 6). Monitor. https://www.monitor.co.ug/uganda/magazines/farming/why-you-should-have-a-shelter-for-your-cattle-1724530

Wiki: Open Source Ecology. (n.d.). *Wind turbine - open source ecology.* Wiki.opensourceecology.org. https://wiki.opensourceecology.org/wiki/Wind_Turbine

Willsher, I. (2022, December 12). *The pros and cons of off-grid living.* Utopia. https://utopia.org/guide/the-pros-and-cons-of-off-grid-living/

Wiring your off grid home. (2023, January 3). Backwoods Solar. https://backwoodssolar.com/learning-center/living-off-grid-articles/learning-center-wiring-your-off-grid-home/